T0138617

Mobile Apps Engineering
Design, Development, Security, and Testing

Mobile Apps Engineering
Design, Development, Security, and Testing

Edited by
Ghita Kouadri Mostefaoui
Faisal Tariq

CRC Press
Taylor & Francis Group
Boca Raton London New York

CRC Press is an imprint of the
Taylor & Francis Group, an **informa** business

A CHAPMAN & HALL BOOK

CRC Press
Taylor & Francis Group
6000 Broken Sound Parkway NW, Suite 300
Boca Raton, FL 33487-2742

© 2019 by Taylor & Francis Group, LLC
CRC Press is an imprint of Taylor & Francis Group, an Informa business

No claim to original U.S. Government works

Printed on acid-free paper

International Standard Book Number-13: 978-1-138-05435-6 (Hardback)

Visit the Taylor & Francis Web site at
http://www.taylorandfrancis.com

and the CRC Press Web site at
http://www.crcpress.com

Contents

Preface

MOBILE COMPUTING HAS CHANGED the way we learn, interact with online services, and manage information. The popularity of handheld devices among people of all ages and cultures has increased the demand for highly interactive and user-friendly mobile apps. The multitude of sensors available in mobile devices such as GPS, ambient light sensing, and accelerometers have broadened the use of mobile apps in various application domains. Mobile apps vary widely, from weather forecasting and managing a patient's health to providing online education, among many others.

Both students and lecturers of software engineering with a particular focus on mobile app development struggle to find a self-contained guide on how to follow the development life cycle of a mobile app project. In the great majority of these projects, the process generally follows a traditional software development life cycle—namely, setting up a set of requirements and then following an incremental development of the mobile app up to the achievement of acceptable functionality and design. A mobile app is, however, very different from a desktop application. For instance, mobile apps are expected to run on multiple mobile operating systems, various screen sizes, and diverse technologies. Testing of mobile apps is therefore different from that of desktop applications. Additionally, mobile apps differ in their context of use and may need to take a number of factors into consideration including internet connection availability and speed, computational complexity, memory requirements, battery status, and accessibility features. These factors affect the software life cycle of a mobile app project and therefore more suitable architectures, design patterns, and testing approaches are needed. In practice, students as well as developers use their experience in desktop application development and customize the methodologies and tools to fit the particularities of a mobile app. We believe that a more structured approach can supplant this ad hoc one.

The objective of this edited book is to gather best practices in the development and management of mobile apps projects. We aim at providing software engineering lecturers, students, and researchers of mobile computing a starting point for developing successful mobile apps. To achieve these objectives, we emphasize the essential concepts such as app design, testing, and security. This has been done with the aim of producing a compact yet self-contained book to stimulate further research interest in the topic. We believe our effort can make mobile apps engineering an independent discipline inspired by traditional software engineering, but taking into account the new challenges posed by mobile computing.

Acknowledgments

W E WOULD LIKE TO thank Dr. Mitul Shukla for his support in the early stages of the current project.

Editors

Ghita Kouadri Mostefaoui is a Senior Teaching Fellow at the Department of Computer Science, University College London, UK. Previously, she worked as a software engineer at the UK's synchrotron. Prior to that, she was a researcher at Oxford University. Ghita is primarily interested in best practices in teaching/learning computer science in higher education. She is also interested in mobile apps technologies.

Faisal Tariq is a Senior Lecturer in the School of Engineering at the University of Glasgow, UK. Prior to this, he worked at Queen Mary University of London and the University of Bedfordshire. He is also serving as editor of *Journal of Networks and Computer Applications*. His research interests include future wireless networks, technologies for smart cities, and mobile computing and communications.

Contributors

Junaid Arshad
School of Computing and Engineering
University of West London
London, United Kingdom

Silvia M. Ascate
Institute of Computing
Federal University of Amazonas
Manaus, Amazonas, Brazil

Pedro Borges
Department of Informatics Engineering
University of Coimbra
Coimbra, Portugal

Michael Chai
School of Electronic Engineering
    and Computer Science
Queen Mary University of London
London, United Kingdom

Tiago Cruz
Department of Informatics
    Engineering
University of Coimbra
Coimbra, Portugal

Arilo Dias-Neto
Institute of Computing
Federal University of Amazonas
Manaus, Amazonas, Brazil

Michael Guckert
KITE – Kompetenzzentrum für
    Informationstechnologie
Technische Hochschule Mittelhessen
Friedberg, Germany

Han Jing
Department of Computer Science
University of Southern California
Los Angeles, California

Patty Kostkova
Institute for Risk & Disaster Reduction
University College London
London, United Kingdom

António Lima
Department of Informatics Engineering
University of Coimbra
Coimbra, Portugal

Jonathan Loo
School of Computing and Engineering
University of West London
London, United Kingdom

Andreea Molnar
School of Software and Electrical
    Engineering
Swinburne University of Technology
Melbourne, Victoria, Australia

Eduardo Noronha de Andrade Freitas
Departament of Informatic
Federal Institute of Goiás
Goiânia, Goiás, Brazil

Kariny Oliveira
Institute of Computing
Federal University of Amazonas
Manaus, Amazonas, Brazil

Paulo Simões
Department of Informatics Engineering
University of Coimbra
Coimbra, Portugal

Bruno Sousa
One Source Consultoria Informática Lda.
Coimbra, Portugal

Gabriele Taentzer
Faculty of Mathematics and Computer
  Science
Philipps-Universität
Marburg, Germany

Steffen Vaupel
Faculty of Mathematics and Computer
  Science
Philipps-Universität
Marburg, Germany

Isabel K. Villanes
Institute of Computing
Federal University of Amazonas
Manaus, Amazonas, Brazil

CHAPTER 1

An Introduction to Mobile Device Security

António Lima, Pedro Borges, Bruno Sousa,
Paulo Simões, and Tiago Cruz

CONTENTS

1.1 INTRODUCTION

This chapter provides a state-of-the-art overview regarding available solutions and best practices for mobile device and mobile application security. It provides a broad perspective of the security aspects involved in mobile applications, encompassing the whole ecosystem (user, operator, operating system, mobile application, and user and application data). Besides providing guidelines and best practices for the development of secure mobile applications, it also includes information on how to apply such solutions in Android and iOS mobile operating systems.

This chapter includes several techniques that can be applied by different players, including application developers (to design and develop secure mobile applications) and administrators of corporate systems (to enforce security policies on mobile devices). The myriad of uses for mobile devices raises security threats that require solutions involving the full chain: device manufactures (e.g., Samsung), services providers (e.g., Google with Play Store), applications developers, and consumers (end-users). As such, a broad perspective is required encompassing the design and development of secure mobile applications, including selection of appropriate platforms (operating systems, mobile devices, telco support, mobile device management frameworks), best practices for developing secure mobile applications, and a clear understanding of the different stakeholders' roles in provisioning secure mobile applications. For instance, consumers need to understand the risks associated with different uses of their mobile device(s), and corporations need to provide adequate security mechanisms for their devices or for the devices owned by consumers but employed for professional activities—in line with the Bring Your Own Device (BYOD) paradigm, which extends the possibilities of mobile device use even in mission-critical scenarios [33], but leads to issues regarding the management of security by corporations.

Several security approaches can be pursued to prevent attacks (e.g., installation of malware), to monitor the behavior of applications and use of mobile devices (e.g., applications sending frequent data to remote sites without user permission) and to mitigate potential threats (e.g., malware accessing personal data). In particular, the Mobile Device and Application Management (MDAM) frameworks (e.g., Samsung Know, Android for Work) are essential tools for corporate environments, to enforce security policies.

1.2 SECURITY RISKS AND OPEN CHALLENGES

The increased computing, communications, and interface capabilities have allowed mobile devices to progressively evolve towards becoming diversified digital assistants. Nevertheless, such newly acquired roles have also shown themselves to be a hindrance in terms of security as they have increased the value of the mobile device as a target for attackers.

The motivation for this chapter is many-fold. First, to bring users awareness of the security risks associated with the use of the mobile devices. Second, to provide an introduction to the security mechanisms that can be employed to prevent and mitigate security threats. Third, to highlight how corporations can manage devices and applications running in corporate devices or even accommodate new trends such as Bring Your Own Device, especially in mission critical scenarios such as public protection and disaster relief where security requirements are more stringent. Fourth, to discuss how device manufacturers can design and incorporate security mechanisms that support trusted applications in mobile devices, such as the trusted execution environment.

The evolution of mobile devices requires security concerns in all the layers of the chain associated with mobile devices, from the design phase of device manufactures until the end of the chain involving consumers and their behaviors. The insecure usage of a device exposes the information contained in it to attackers, who can be quite resourceful by employing several techniques to defeat existing security measures.

1.2.1 Evolution of Mobile Device Complexity

Contemporary mobile devices are capable of performing tasks that were not envisaged to be feasible or even possible a few years ago. This is due to an evolutionary process that ultimately brought together the results of advances in several areas such as integrated electronics, batteries, communications, interface technologies, and software design.

Considering the importance of such developments, this subsection provides an overview of the evolution of mobile devices and how their increasing complexity allows users to perform more complex and demanding tasks while also posing significant challenges for manufacturers and application developers to build secured and trusted applications and devices.

The notion of mobile devices, nowadays, is associated with devices that do not present any restriction to the movement of the carrier or host. By facilitating the mobility of people, these devices are characterized by several traits, among which are dimensions and weight (e.g., can be lost easily), connectivity to other devices (e.g., can communicate with other devices without user awareness), and when the device mobility occurs (e.g., can connect to insecure networks).

Most mobile devices have an embedded display screen and receive input either from physical buttons and keyboards or from on-screen virtual touch-buttons and keyboards. Some devices also support audio input (e.g., voice recognition) and sensors (e.g., accelerometers, compasses, magnetometers, and gyroscopes allowing the detection of orientation and motion). Data capture devices such as barcode, RFID, fingerprint and smart card readers can also be embedded or attached to these mobile devices. On top of all the already mentioned features, connectivity usually falls under 3G/4G, Wi-Fi, Bluetooth, NFC, and GPS capabilities.

The first mobile devices were poor regarding supported features. Devices such as the IBM Simon (released in 1994, regarded as the first smartphone) [35] were very "dumb" by today's standards, having very limited electronics and processing power. Table 1.1 compares mobile devices (almost) equidistant in time: an iPhone X, an iPhone 6, an iPhone 1st Generation (commonly referred to as the first actual smartphone), and the IBM Simon. Mobile devices such as smartphones have come a long way, with dramatically increased computing power and capabilities. Given the plethora of possibilities that these more modern mobile devices offer, it is no wonder that they have spread to almost every job that is associated with mobility and have even begun to substitute previously dominant (and often unportable) forms of computing devices such as desktop computers.

1.2.2 Information Risks

Many mobile device users are not aware that the behaviors with which they engage and use the devices may put the security of the devices and their information at risk (e.g., credit card details or biometric data used for authentication). Users engage in risky security behaviors by pursuing "always connected" paradigms, disregarding the security supported by Wi-Fi networks, and the "uncontrolled" installation of applications, even from official application stores. Such security risk behaviors are highly correlated with a defective security culture. First, users have poor security and privacy awareness. Second, users lack knowledge of prevention and protection mechanisms to secure information in their devices. Security studies [2,36] evidence the risky behaviors that users adopt. For instance, users believe that applications in the official application stores are safe, but this assumption is not correct as, despite the security control in such stores, applications with malware that could be installed by everyone have been found.

Users get engaged in a vast range of risky behaviours. Android uses a permission system to protect its vulnerable components, but many times users do not pay attention to the permissions that are being requested or do not even understand the purpose of the permissions. Most users do not refrain from accessing public unsecured Wi-Fi networks, and almost half of the users in security studies had never even thought of the security risk that accessing an unsecured Wi-Fi network poses. One particularly risky behavior that users engage on is chatting with strangers. Unknowingly, users are subjecting themselves to manipulation or deception that may lead them to perform actions that can allow the attacker to enter or access the information on the mobile device. Storing highly sensitive photographs on the mobile devices, especially in the default gallery applications without any kind of protection, is also dangerous. This allows an attacker that has access to the device to extract them easily and use them for monetary profit, either by selling them or by asking for a ransom. Location services are another key point on privacy to which users do not pay attention. By having location services turned on all the time and by sharing their

TABLE 1.1 Evolution of Mobile Device (Smartphone) Specifications

	IBM Simon	iPhone (1st Gen.)	iPhone 6	iPhone X
Release Date	August 16, 1994	June 29, 2007	Sep. 19, 2014	Nov. 3, 2017
Dimensions	200 × 64 × 38 mm	115 × 61 × 11.6 mm	138.1 × 67 × 6.9 mm	143.6 × 70.9 × 7.7 mm
Weight	510 g	135 g	129 g	174 g
CPU	16-bit, 16 MHz, x86-compatible	32-bit, 412–620 MHz, ARM	64-bit, 1.4 GHz dual-core, ARM	2.39 GHz hexa-core 64-bit
GPU	None	PowerVR MBX Lite 3D GPU	PowerVR Series 6 GX6450 (4-core)	Apple GPU (3-core)
Memory	1 MB	128 MB	1 GB LPDDR3	3 GB LPDDR4X
Storage	1 MB	4, 8, or 16 GB	16, 64, or 128 GB	64 or 256 GB
Battery	7.5 V NiCad	3.7 V 1400 mAh Lithium-ion	3.82 V 6.91 Wh (1,810 mAh) Lithium-ion	3.81 V 10.35 Wh (2716 mAh) Lithium-ion
Data Inputs	Microphone, touchscreen with stylus	Multitouch touchscreen, 3-axis accelerometer, Proximity sensor, Ambient light sensor, Microphone	Multitouch touchscreen, microphone, gyroscope, accelerometer, digital compass, iBeacon, proximity sensor, ambient light, fingerprint, barometer	Multitouch touchscreen, microphone, gyroscope, accelerometer, digital compass, iBeacon, proximity, ambient light, fingerprint, barometer, face ID
Display	4.5 in × 1.4 in, 160 px × 293 px monochrome backlit LCD	3.5 in diagonal, 320 × 480 px resolution at 163 ppi, 2:3 aspect ratio, 18-bit (262, 144-color) LCD	4.7 in diagonal, 1334 × 750 px resolution at 326 ppi, LED-backlit IPS LCD, 16:9 aspect ratio	5.8 in diagonal, 2436 × 1125 px resolution at 458 ppi, Super Retina HD: AMOLED, 19.5:9 aspect ratio
Camera	None	Rear: 2 MP	Rear: 8 MP, 1080p HD video recording, slow-motion video, panorama; Front: 1.2 MP, 720p video recording	Rear: 12 MP, quad-LED dual-tone flash, slow-motion video, panorama, face recognition, 1080p and 4K video recording; Front: 7 MP, 1080p HD video, face detection
Connectivity	2400-bps Hayes-compatible modem, 9600-bps Group 3 send-and-receive fax, I/O connection port, PCMCIA type 2	Quad-band GSM/GPRS/EDGE, Wi-Fi (802.11 b/g), Bluetooth 2.0	UMTS HSPA + DC-HSDPA, CDMA EV-DO Rev. A and Rev. B, GSM/EDGE, Wi-Fi, Bluetooth 4.2, NFC, GPS & GLONASS	LTE, TD-LTE, UMTS, HSPA+, DC-HSDPA, GSM/EDGE, Wi-Fi, Bluetooth 5.0, NFC, GPS, GLONASS, Galileo & QZSS

location on social networks, users expose themselves to attackers that may want to harm them physically or benefit from their being in a specific location allowing them to engage in burglary, robbery, or theft.

While risky behaviors by end-users may to some extent only affect themselves, in corporate environments the consequences might affect working teams and organizations. As such, these attitudes and behaviors are of even higher importance in such environments. By having a security leak and allowing any sensitive information to fall onto the hands of attackers, whole work teams can be disrupted—resulting in elevated costs for the company. For instance, a company devoted to fraud detection in credit card transactions might simply lose its customers and never recover their established reputation.

Additionally, because of the attachment employees have to their personal mobile devices, organizations are pressured to allow BYOD approaches. While this boosts productivity and employee satisfaction [22], it also poses serious security risks to organizations. In the corporate environment, it has been found that either enabling security features by default on the mobile devices or educating users toward engaging in security-enforcing practices can enhance security in the enterprise environment [13,29,32].

1.2.3 Threat Taxonomy

For a better understanding of the risks presented in the previous section, it is useful to identify the most common strategies employed by attackers to infiltrate malicious code onto mobile devices. Attackers commonly try to exploit either technical vulnerabilities in the operating systems or a person's naivety or goodwill in performing actions for them in order to achieve infection. Malicious software can make its way into mobile devices in several ways:

Social engineering. Users are manipulated to install or run malicious code unwittingly. This is done in several ways: application repackaging, where malicious code is included in a legitimate application; fake installers that mimic known application installers; malicious updates, where a legitimate application is released, but afterwards malicious code is deployed via updates; and drive-by download, where the user browses a web page that automatically downloads malicious code.

SMS/MMS. In this type of attack, flaws in the design of the SMS architecture are exploited and can be used to deliver malicious code or to maintain communication with the attacker. The malicious code in SMS/MMS messages can make the device disconnect from the network, terminate calls, reboot, and crash [27].

Bluetooth. Attack by this route exploits this form of radio communication to spread malicious code from one device to another [31]. A compromised device can pair itself with other devices using default passwords, and then send copies of malicious code to the target device.

Internet access. Devices can connect to the internet using several technologies such as Wi-Fi or 3G/4G. This enables them to use services such as email, social networks, or web browsing, but also makes them subject to the same threats as conventional personal computers. Furthermore, the devices are switched on most of the time, which combined with a constant internet connection increases the probability of an attack.

USB propagation. This sort of attack exploits the ability for application installation via a data cable. For instance on Android devices, by using the adb shell, the permission system can be bypassed, making it possible for a malicious application to be installed with access to all the permissions. Furthermore, there is some malware that infects Windows machines and installs adb command line tools that propagate malicious software to the device through USB when it is connected [37].

Near-field communication (NFC). This kind of attack exploits NFC, which can be used in a drive-by type of attack to force the Android device to browse a website that executes malicious code on the device. Furthermore, an Android device with an NFC-capable malware may interact with contactless payment cards in the vicinity and perform malignant transactions [47].

1.2.4 Case Studies on Security Threats in Mobile Devices

This section complements the previous discussion with the presentation of a few case studies regarding security threats in mobile devices.

There are several reports documenting the participation of mobile devices in distributed denial of service (DDoS) attacks. In particular, CloudFlare presented in 2015 a DDoS report [18] in which more than 70% of the DDoS sources were mobile devices. The analysis shows that users on mobile devices were victims of *ad networks*, where malicious JavaScript was used to launch a flood of XHR (XMLHttpRequest) requests against CloudFlare servers. Other reports confirm the use of mobile devices as active components in DDoS attacks. Radware [39] states that, due to their computing capabilities and to the availability of various exploits, bots, and remote access tools in app stores, mobile devices are easily controlled to carry out DDoS attacks.

Other cases studies, which seem to be more frequent, are related to malware or spyware applications, which intend to exfiltrate users' personal data. For instance, the Lipizzan spyware [9], detected by Google in the Android security program, is able to perform call recording, perform location monitoring, take screenshots, and fetch user information (e.g., contacts), among other features. Indeed, the Lipizzan spyware employed impersonating features to avoid detection by Google Play security mechanisms. For instance, Lipizzan announced itself as being a backup application, while in reality after installation it would download malicious code intended to take control over the mobile device (i.e., rooting the device).

In the iOS ecosystem [21], other examples of exfiltration have also been reported, which do not require the installation of applications, but rather the pairing of a device running iOS and a computer with iTunes software. Such attacks can lead to the exfiltration of personal data such as the address book, SMS massages, notes, calendar, and call history. This kind of attack, despite requiring the pairing of the iOS device with a computer, can be performed without being noticed even in corporate networks, where antivirus and antimalware software can be in place.

Despite the existence of those case studies and other reported incidents, it should be noted that the number of mobile security breaches documented in security reports is still quite low compared to the number of occurrences [40]. Indeed, a high percentage of IT security professionals state that it has become harder to guarantee the security of mobile devices. The attacks range from malware to key logging, credential theft, and phishing attacks.

1.3 FUNDAMENTAL CONCEPTS

This section provides an overview of the fundamental concepts employed to attain a secure environment on a mobile device. It describes the concepts of device security and application security, the different goals in consumer vs. corporate environments, the role of the stakeholders in achieving and maintaining a secure environment, the management of the devices and associated assets, and finally some methods that can be used to keep a device and its applications secure.

The concepts are detailed for a single operating system (OS) to ease comprehension of the underlying paradigms; as such, the Android OS was chosen due to its popularity and its openness in comparison with iOS.

1.3.1 Device Security and Application Security

The Android OS runs on top of a kernel derived from the Long Term Support (LTS) Linux releases. This is due to the Linux kernel having a robust driver model, efficient process and memory management, and networking capabilities. It is customized for use in embedded environments with constrained resources such as mobiles devices.

In order to run on the Android OS, applications are compiled into a custom byte-code: the Dalvik EXecutable (DEX) format. This allows each application to be executed in its own Dalvik or Android Runtime (ART) virtual machine instance (depending on the OS version),

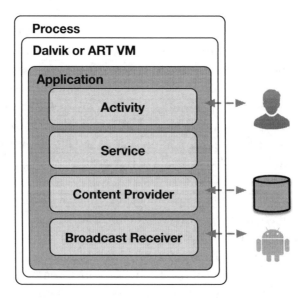

FIGURE 1.1   Android application security.

as pictured in Figure 1.1. Each instance runs with a unique UNIX user identity, so that applications are isolated in the Linux platform subsystem and have their own space in the file system to write private data. This design is enforced at the kernel level by discretionary access control (DAC) and prevents applications and services from disrupting each other. Furthermore, Security Enhanced Android (SEAndroid) is a port of Security Enhanced Linux (SELinux), a Linux kernel security module that supports access control policies such as mandatory access control (MAC) [44]. It was introduced in Android 4.2, in a permissive configuration, and from Android 4.4 onward it is present in an enforcing configuration.

There are four key components of Android applications that are important for understanding the behavior of applications and the security architecture of this OS:

Activities represent the way the user interacts with the application. Each different screen of the application is defined by a different independent activity, even though they work together to provide a consistent user experience. Activities are launched using *intents* and can be started by a different application, if it is configured as allowed.

Services are components without any type of user interface that perform long-term background tasks or work for remote processes. Services are launched using *intents* and other components. For instance, activities can start services to do some work or bind to them for interaction.

Content providers manage shared sets of data for access from within the application or between different applications. The data can be stored in the file system, in SQLite databases, in the web, or in any other persistent storage. The content provider enables the applications to query or even modify the data if the configured permissions allow it. Content providers are accessed through the application-defined Uniform Resource Identifiers (URIs).

Broadcast receivers are components without a user interface that listen to system-wide broadcast announcements that perform minimal amounts of work or launch services. Broadcasts can be initiated by the system or by application-defined events. For example, they can answer to system events like BATTERY_LOW, BOOT_COMPLETED or SCREEN_ON.

Intents work as the main vehicles of intercomponent communication (ICC), since applications are isolated in their own sandbox. *Intents* are abstract representations of actions to be performed and can be used to launch *activities* and to communicate with components of the same application such as *broadcast receivers* and *services*. This is performed through the binder interprocess communication (IPC), an IPC framework.

ICC is mediated by the Android middleware and obeys the permissions established in the manifest file. It is the developers' responsibility to specify the required permissions in the manifest file. A component protected by a set of permissions can only be accessed by other components to which access has been granted.

In the Android OS the sensitive hardware components (e.g., telephony, camera, network) are protected behind a permission model to avoid unauthorized access that could interfere with the system performance and the user experience. Therefore, each application that wants to access a particular protected part of the system must require such permission in the manifest. During the installation, the system verifies the authorities that signed the application's certificates and can also inquire of the user whether to allow or deny permission. If such permission is granted the application is able to make use of the protected resources. If the permission is not granted the application will work but the access to the denied resources will not work and no notifications will be shown to the user. Permissions can be grouped into four types, according to the risk associated with granting access [3]:

1. `Normal`, the default low-risk permission that gives access to application-level features with minimal risk to the system, applications, or user. It is automatically granted by the system without the user's approval.

2. `Dangerous`, a high-risk permission that enables the application to control the device or to access private data. The system does not automatically grant it to the requesting application. Instead, the dangerous permissions are presented to the user for approval.

3. `Signed`, a permission that is granted by the system to an application that has the same certificate as the application that declared the permission. If the certificates match, the permission is automatically granted by the system.

4. `Signature Or System`, a permission that is granted by the system only to applications that are part of the Android system image or to applications with the same certificate as the declarer. This permission is used only on special situations where multiple vendors have applications that need to share resources explicitly while being built together.

Despite all these features aiming to protect applications and their data from the rest of the device, attackers still find ways to interfere with the data. In this sense, for the purpose of adding an extra layer of security, trusted execution environments (TEE) can be employed (further details are described in Section 1.5.7). TEEs can be used to perform critical operations that cannot be hijacked or have their data modified by external parties. For instance, specific ARM processors with the TrustZone [7] technology have dedicated hardware functionality to perform these sensitive operations. This allows vendors to develop applications that have enhanced security features backed by hardware isolation.

1.3.2 Consumer vs. Corporate Environments

There are different security requirements for consumer and corporate environments. Consumer security is more focused on threats and attacks related to personal information leakage: bank account credentials; credit card fraud; monetary extortion; biometric data; private images and messages, etc. On the other hand, in corporate environments, where mobile devices are used to perform business or organization-specific tasks, security is more focused on preventing the leakage of confidential information and malicious actions that can hinder the use of devices. For instance, in specialized fields such as public protection

TABLE 1.2   BYOD Schemes and Enterprise Mobility Schemes

BYOD Scheme	Device Control	# Devices in Corporate
Corporate Owned Corporate Enabled (COCE)	++++	+
Corporate Owned Personally Enabled (COPE)	+++	++
Personally Owned Corporate Enabled (POCE)	++	+++
Personally Owned Personally Enabled (POPE)	+	++++

and disaster relief (PPDR), the security requirements concern the leakage of critical mission information, the location of the agents in the field, and attacks on the device that make it unusable and unable to perform its function [33].

When using paradigms such as the already mentioned Bring Your Own Device (BYOD), single devices may need to support multiple profiles. BYOD schemes can be classified regarding device ownership and enabling model, as summarized in Table 1.2. The different types of BYOD schemes available for enterprise usage introduce distinct mobile device management (MDM) policies, which define admissible devices and the type of controls held over them (accessible data, applications, and functionalities). Recently, large-scale companies have begun to move from the COCE scheme to more BYOD friendly alternatives, such as COPE and POCE [1].

General concerns relating to the limited battery life inherent in mobile devices apply to both consumer and corporate environments. Mobile devices' compactness also poses concerns when it comes to cooling or increasing storage space. Another concern, often treated separately, is the risk of misplacement or theft of mobile devices—which allow completely different suites of attacks or paths to the data stored in the device.

1.3.3 Stakeholders

Figure 1.2 provides a simplified view of the stakeholders of mobile device ecosystems. For sake of simplicity some relationships have been omitted (e.g., relations between service providers and developers, or between consumers and corporations).

Several stakeholders can be identified. Telecommunications and network operators provide network connectivity to mobile devices, often having contractual relationships with the consumer or corporations. Device vendors manufacture and provide devices and are responsible for warranty processes and firmware updates. Device vendors may be interested in adding new security hardening features to their devices to attract more customers, either by customizing the OS or by adding extra hardware security layers. Service providers provide services to the users or corporations and may require specific applications in the mobile device (take the example of Netflix or others). The operating system's developers, such as Google or Android are responsible for building, maintaining, and supporting the operating system, releasing regular updates. Such companies also provide a plethora of services that are managed through specific platforms like the official application stores. Indeed, such stores are crucial for spreading updates of applications. OS developers obviously, are always striving to improve the quality of their product, which implies improving the security. The cost of security flaws in the OS is a reduction of user trust in the OS. Developers are also

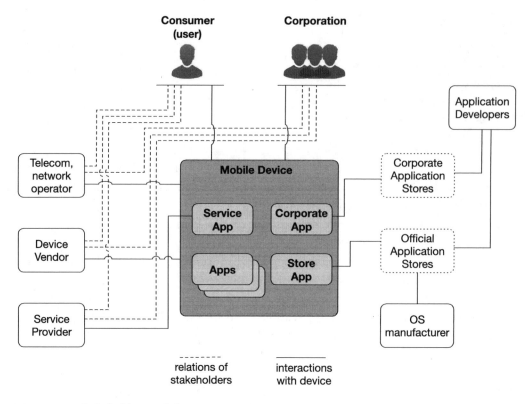

FIGURE 1.2 Stakeholders and their interactions.

important key players in this ecosystem, by developing applications and releasing them in the dedicated application stores. All of these stakeholders take part in shaping how the ecosystem is maintained and developed.

From a user perspective, a secure mobile device means that the user can use it in a more relaxed way which will improve the relationship with the given device or OS. On the other hand, corporations can manage their own application stores to enforce security policies in their own devices, or to manage to some extent the applications that run on personal devices (e.g., BYOD devices as described in Section 1.3.2).

1.3.4 Mobile Device and Application Management

Mobile device and application management (MDAM) frameworks provide the means for an entity to maintain control of all its managed mobile devices while providing protection and enhanced security to the data, in order to reduce information risks (recall Section 1.2.2). While these software suite managers are able to deploy configurations, applications, and policies across several mobile devices swiftly, specially in corporate environments, they can also be used to remotely control the managed devices and enforce access restriction mechanisms, in case a security threat is detected.

MDAM solutions in conjunction with systems such as network access control (NAC) are key enablers of the BYOD paradigm [19]. Furthermore, by coupling MDAM solutions with mobile forensics or intrusion detection systems (IDS), mobile device usage in the workplace is enhanced by assuring that all the accesses to sensitive information are monitored and evidence is collected regarding unauthorized disclosure of information or other kinds of attacks.

Further details on available MDAM solutions can be found on Section 1.4.

1.3.5 Security Life Cycle for Mobile Devices

The security life cycle for mobile devices encompasses three main phases, as illustrated in Figure 1.3:

1. *Prevention*, which mainly focuses on what can be avoided or stopped before it becomes a threat or compromises a device.

2. *Monitoring and detection*, which are focused on the constant and methodical analysis of device and application behaviors.

3. *Reaction and/or mitigation*, which are concerned with containing or stopping a threat, even if it has already caused damage.

The guidelines provided by NIST for Intrusion Detection and Prevention [45] fit within this context. The following sections detail the most common prevention, monitoring, and mitigation methods found in mobile devices along with their respective advantages and disadvantages.

1.3.5.1 Prevention

The preventive methods are gaining more emphasis for providing data security on mobile devices, when compared with monitoring or detection approaches. Prevention is the first phase of the security life cycle and is also less intrusive, avoiding waste of resources such as power consumption in mobile phones. The takeaway message from this section is that albeit both defensive strategies can be effective on their own, they are not mutually exclusive and in fact may benefit vastly (in terms of effectiveness) from being combined.

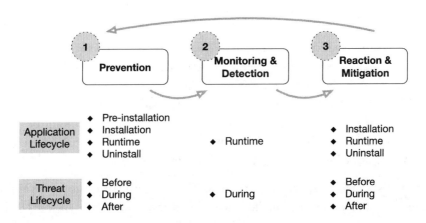

FIGURE 1.3   Security, application, and threats life cycle(s).

Several prevention methods have been adopted as a precautionary measures in mobile devices to minimize the amount of potentially harmful malware applications that get into the device or that might exploit known attack vectors. An attack vector is a method or exploited security feature used to compromise the security of a system. Most commonly used prevention methods include the following:

Sandboxing. Creation of separate virtual spaces for the untrusted applications or code, in order to run with limitations so as to diminish possible attack vectors: applications run with no interactions with other applications and under tight restrictions in the underlying host system. Most modern mobile device operating systems, such as Android, iOS, and Windows Phone, run their applications in sandboxed environments.

Personalized approval. Detailed analysis of each application's conformity to security policies before being approved for distribution. The most common example of this technique is Apple's App Review system wherein each submitted application goes through a series of meticulous review steps before becoming available in the App Store. Microsoft has a similar system for the Windows Phone App Store. By contrast, Google has no similar policies for applications available in the Google Play Store.

Code and asset assessment. Well-known types of malware can be identified by distinct signatures. Most antivirus protection mechanisms rely on signature databases to identify potential threats to the system. Another way of assessing applications is to check which system calls, frameworks, and methods they use and under which circumstances. Both iOS and Windows Phone enforce this behavior through their official application stores, each submitted application for which is inspected before becoming available.

Compartmentalization. The basic premise is to completely separate the user's personal space from the work space within the device, avoiding unnecessary compromises and encapsulating threats from the personal space. The work space is usually subject to stricter security restrictions and might even not allow installing or running applications not allowed by security policies. Android devices can have user accounts that are managed by the main account, which can have a more limited set of permissions.

Trusted party management. This method relies on the verification of trusted certificates, keys and signatures to manage which entities can interact with the system or which applications can be installed. For instance, only a single trusted computer might be able to write to the mobile device's storage (aside from the device itself) or only apps approved and properly signed by a company beforehand might be installed on the device. Android, iOS, and Windows Phone all have the capability to install and manage custom trusted enterprise applications in the devices.

Permission systems. One way of circumventing the ordeal of assessing applications' functionalities or used tools is to leave it to the user to either accept or reject a list of granted permissions. Android and iPhone apps, for instance, must explicitly detail which system features or hardware capabilities they (might) need access to. For Android devices, this is a mandatory step performed before the app is installed in the device, while for iPhone devices each individual permission will be asked during runtime when needed for the first time. In both scenarios, this information can be checked before installing the application and it is up to the user to gauge the risk when granting access to sensitive features, such as access to the camera or SD storage.

Authentication schemes. These comprise the use of encryption standards, with the possibility of depending on the existence of an external super user. The idea is to lock system management behind complex password/PIN mechanisms before any (or even every) write or read action is performed on the system. Most modern mobile operating systems come with the option to encrypt the phone regarding a locking mechanism (usually a PIN code or a fingerprint) without which the contents of the phone cannot be accessed.

Limited time access. Ephemeral fixed-length sessions are scheduled for the device's usage, and all services are interrupted or aborted when the allocated time slot elapses. Not available by default, this method is usually implemented at application level by using timed access tokens.

1.3.5.2 Monitoring

Monitoring is an important component of the security life cycle, enabling detection and profiling of malicious activities within the system, during runtime. Most common approaches include the following:

Behavior analysis. Focused on what is and is not expected of each type of application, the application or code is scrutinized during runtime for any anomalous or suspicious activity, for detecting unwanted accesses or transfers of delicate or out-of-scope data. This process may be coupled with a training phase or expected default behavior thresholds for known applications [11,14].

Log analysis. The host system logs systematic information regarding memory and battery usage, storage access, data transfers, and used services (e.g., Bluetooth, Wi-Fi, mobile data, etc.) to portray the app's true nature. The logs are periodically sent to a trusted supervising server to delegate the heavier, more resource-intensive, analysis burden out of the mobile device, albeit some minor analysis can still be performed on site. Log integrity measures should be considered on both end points for this method to work properly [8,30,34].

System call hooking (SCH). The act of altering the underlying host system for more fine-grained logging of the system's function calls and frameworks used (along with other features) with the goal of easing the monitoring task [15]. This can help detecting unnecessary use of mobile device resources such as cameras or microphones.

Runtime permission checks. Every hardware and software feature external to the application requires explicit permissions for said actions to be granted at runtime, at the host system's level. One common example would be requesting permission to use the device's camera every time to stop a malicious application from stealthily collecting unwanted pictures of the user or their surroundings. Android and iOS have runtime permission verifications, allowing for the user to toggle or opt out of permissions after installing the application.

The main advantages of monitoring security strategies include a better and more refined assessment of the sensitive data and what actions or applications could be compromising it, hopefully in time for intervention. They also enable a stricter control of what are acceptable behaviors, through detailed logs and variable limitations. Given the limited computing power and battery life, some of these methods can easily become an overkill for the device or rely too heavily on consistent and systematic connection with secure networks and servers, which might not be a possibility, depending on the use case scenario. Another disadvantage is the risk of potential interventions in "real time" which, depending on network status, system resources, cryptographic complexity, and connectivity issues, might result in unacceptable latency.

It is worth mentioning that some monitoring methods cannot be implemented out of the box. A good example is SCH, which is very hard or even impossible to perform on most mobile operating systems without rooting or running a custom modified version of the original OS. Android is the most commonly adopted OS for such barred monitoring techniques, because it is based on the Android Open Source Project and can be altered at core level to accommodate the desired changes more easily.

1.3.5.3 Mitigation

Some of the preventive and monitoring techniques already described in the previous sections can be considered attack vector or vulnerability mitigation techniques on their own, but there are other means to repair or minimize the damage on compromised devices. For instance,

Security policies. Detailing which devices may connect to an infrastructure, what types of data they may contain, what constitutes sensitive data and which operating systems and applications are accepted, in terms of acceptable use and penalties for misuse.

Patch support. The ability to submit updates or patches for an application or system without having to physically access it can be sufficient to fix newly discovered vulnerabilities or potential attack vectors.

Long-term support (LTS). After distribution of the system/application there is a time frame for which the developer agrees to provide support for potential threats.

Remote control. If a device becomes compromised it can be wiped, reset, shutdown, or locked remotely, removing the possibility for an escalated compromise or continued/recurrent access to sensitive data.

Location tracking. If a mobile device is stolen and ends up in a conspicuous location it can be tracked down or simply located after the theft is reported.

Full/coordinated disclosure. If a researcher/developer finds a vulnerability in an application/system, they should contact the developer about their findings and, if given permission (in case of coordinated disclosure), make them available, or simply make them public (in case of full disclosure). This mitigation strategy is controversial as disclosing vulnerabilities can in fact potentiate the discovery of related but stealthier vulnerabilities.

Recall for analysis. The potentially targeted devices are periodically recalled for a temporary system and application assessment. This method can be invasive with respect to personal data if the device is also used for personal purposes or outside of the enterprise environment.

User feedback. A means for the user of the device to report any suspicious or anomalous behavior, no matter how ridiculous it might seem. This must be enforced with a sense of trust between the user and coordinators/feedback-receivers.

Documentation. Documents for scrutinizing successful and unsuccessful aspects of a project, application, or system, along with identified threats, how they were handled, and how they could have been avoided. This process is usually performed at the end or at predefined iterative stages of the life cycle, a post mortem document is a common example of this strategy.

1.3.5.4 Analysis and Comparison

The security life cycle includes several phases and security approaches (see Figure 1.3)—prevention, monitoring, and mitigation/reaction—but not all of them are equally effective or suited to the same situation: context is key. This section summarizes and compares these security approaches.

Table 1.3 analyzes at which moment of an application's life cycle the already mentioned prevention strategies act, and for which mobile OSs (Android, iOS, and Windows Mobile) they can be implemented or are already available. Many prevention strategies occur at runtime, as opposed to the mechanisms held in stationary devices, such as desktop

TABLE 1.3 Key Moment of Action per Prevention Method

	Pre-Installation	Installation	Runtime	OS Availability
Sandboxing			•	Android, iOS, Windows Mobile
Personalized Approval	•			iOS, Windows Mobile
Code and Assets Assessment	•			iOS, Windows Mobile
Compartmentalization			•	Android
Trusted Party Management		•		Android, iOS, Windows Mobile
Permission Systems		•		Android, iOS, Windows Mobile
Authentication Schemes			•	Android, iOS, Windows Mobile
Limited Time Access			•	Android, iOS, Windows Mobile

TABLE 1.4   Prevented Threats per Monitoring Method

	Privilege Escalation	Worms, Trojans, Viruses	Resource or Data Misuse	Spyware	Theft, Device Cloning	OS Availability
Behavior Analysis		•	•	•	•	Android, iOS, Windows Phone
Log Analysis		•	•	•	•	Android, iOS, Windows Phone
System Call Hooking	•	•	•	•		Android
Runtime Permission Checks			•			Android, iOS, Windows Phone

computers. This is because it can be harder to find suspicious applications solely on the basis of their appearance rather than on their behavior without the constant monitoring provided by antivirus or similar tools; thus, mobile devices rely on runtime approaches.

Table 1.4 summarizes the threats that can be prevented by monitoring mechanisms. The most defective monitoring mechanism in terms of protection is runtime permission checks; the remaining mechanisms seem equally effective against most threats. However, since SCH is usually not readily available on most mobile devices, and behavior analysis requires a priori training, log analysis seems to be the most effective technique that can be employed out-of-the-box.

Although prevention and monitoring can strongly improve security, there will always be new malware or attack vectors against which they are ineffective. The best examples of are zero day, undisclosed, and undetectable exploits, ranging from target-specific applications such as Windows' Stuxnet to global pandemics such as Android's Stagefright. For such threats, mitigation is a more suitable approach.

Mitigation strategies may be applied in the different phases of threats: before, during, or after it is detected or results in damage. Table 1.5 analyzes the point at which each of the addressed mitigation strategies can help handle a potential threat.

The security life cycle mentioned in the previous sections considered three phases, at which distinct security approaches can be applied. It should be noted that other approaches can be considered, such as the observe-orient-decide-act (OODA) loop [12]. OODA is a common approach used by military, commercial operations and learning processes. The

TABLE 1.5   Timing of Mitigation Methods

	Before	During	After
Security Policies	•		
Patch Support	•		•
Long Term Support		•	•
Remote Control		•	•
Location Tracking			•
Full/Coordinated Disclosure			•
Recall for Analysis	•	•	•
User Feedback		•	
Documentation			•

main advantage of OODA is that it focuses on agility when dealing with a threat in order to mitigate its advances and damage at any given point in time, learning from any form of input and feedback obtained.

OODA considers four distinct phases: **Observe**, which includes the identification of threats (1); **Orient** where security approaches are planned according to the threats identified in the previous phase (2); **Decide** where prevention strategies are put in place (3); and **Act**, which includes for instance the mechanisms mentioned in the monitoring and mitigation phases (4). A clear distinction of the OODA approach is that all the phases provide feedback to the Observe phase, in order to assure that all the identified security threats are correctly identified and addressed in the following phases.

1.4 MOBILE DEVICE AND APPLICATION MANAGEMENT

Mobile device and application management (MDAM) is crucial for the security of enterprises, by allowing the management and the remote deployment of applications, configurations, and security policies on a fleet of mobile devices. Moreover, it also provides several features that aid workers in performing their tasks while keeping the devices and the enterprise system secure.

There are several MDAM frameworks available today on different mobile OSs, such as Samsung KNOX [42], Android for Work [28], Flyve MDM [46], or Apple Business [4]. These solutions are provided mainly (but not only) by device manufacturers and major application stores. The key features of the best known MDAM frameworks are presented below.

Samsung KNOX is a proprietary solution from Samsung, integrating multiple technologies to enable a BYOD paradigm. It leverages on the concept of security container, employed to allow enterprise applications to run in an isolated environment while controlling them through MDM application programming interfaces (APIs). It assures trustiness by supporting a secure boot mechanism and relying on runtime protections that are compliant with ARM TrustZone and SELinux.

Android for Work is a platform that supports several management and profile isolation features that rely on components that are already part of the Android OS. Such features include managed profiles, remote provisioning of certificates and configurations, and remote device control. Profile isolation implies separating the user's personal data from the user's business data through the use of containers. It allows the configuration and enforcement of VPN usage on the work profile such that the work data is secured when being transmitted over the internet.

Flyve MDM offers many common features already present in Android for Work. Nevertheless it also offers, besides the management capabilities, monitoring capabilities. It is an open-source solution that can be configured to the enterprise's needs. Its noteworthy features include the configuration of groups of devices, secure enrollment, file deployment on the devices, remote control of mobile device features (e.g., Bluetooth, camera, GPS, Wi-Fi), logging of geolocation of devices, and partial or total deletion of data from the device in case of loss or theft.

TABLE 1.6 Comparison of Mobile Device and Application Management (MDAM) Frameworks

	Samsung KNOX	Android for Work	Flyve MDM	Apple DEP
Secure Boot	•			
Profile Isolation	•	•		
Deployment of Security Policies	•	•	•	•
Deployment of Apps	•	•	•	•
Dedicated App Store		•	•	
Device Monitoring			•	
Remote Data Wipe	•	•	•	
Secure Execution	•			
Supported OS	Android	Android	Android	iOS

Apple Device Enrollment Program is a part of Apple Deployment Programs supporting device management primarily targeted at configuring devices for work when they are deployed. It does not allow BYOD as it can be only applied to corporate-owned devices. It allows the configuration of settings, applications, and services that the employees can use. It automatically configures the devices as they are switched on for the first time and there is no need for interaction from the user or any kind of enrollment.

Table 1.6 summarizes the functionality and features of these MDAM frameworks. Samsung KNOX has many security features lacked by other MDAM solutions because it is a product focused more on security than management, while the others are clearly management solutions with some security features.

1.4.1 Remote Deployment of Security Policies

The remote deployment of security policies allows administrators to push to the devices of the organization sets of configurations aiming at enforcing the required security level. The security policies can include measures such as lockscreen password requirements, mandatory use of device encryption, automatic updates configuration, cloud backups configuration, web browsing configuration, application installation configuration, gaming configuration, hardware capabilities configuration, and messaging and calls configuration.

Through such security policies, corporations can assure that a device behaves according to the defined security policies. For instance, a mobile device only pairs via Bluetooth to authorized devices in such policies.

1.4.2 Dedicated Application Stores

Dedicated application stores arise from the need to have control over the applications the users install on the mobile devices. These dedicated application stores provide applications deemed necessary for work tasks that have been cleared for use on the devices by the administrators. In this way users can only install applications approved by administrators (at least on the work profiles).

Dedicated application stores, such as Enterprise App Stores, are usually managed and maintained as part of a mobile application management (MAM) policy. Its usage can be part of the security policies deployed in the device, allowing organizations to ensure adequate prevention measures regarding which applications are installed on the device. Such policies may also provide the means to control the distribution and deployment of applications downloaded from the official app stores (which may be proxied/cached by the Enterprise App Store), controlled via white and/or blacklist mechanisms. Overall, it can be considered that dedicated application stores combined with the security policies provide corporations with finer control.

1.4.3 Remote Device Control

Remote device control allows administrators to remotely control devices—for instance, to enforce a security policy. It can, for instance, enforce a password-lock on the device if a user is not using one when such is mandated in the security policy. Also, it allows administrators to perform partial or full wipes of the data on the device in most extreme cases, when an anomaly is detected and there is a risk of critical information leakage.

One may think about additional controls, for instance the remote activation of GPS to enable location services. Such feature is required in mission critical services for PPDR environments.

1.4.4 Remote Application Deployment

Remote application deployment allows administrators to deploy applications on the devices without the need of user interaction. This way they can ensure that the applications required to perform work tasks are installed on the devices at all times. Also, when new applications are required to perform work tasks they can be effortlessly pushed to devices, keeping users up to date. Remote deployment also allows groups of devices to simultaneously receive applications and/or files that contain important information to perform a certain mission. For instance, a PDF file with information regarding safety instructions can be provided.

All these features, when combined, provide the required tools for corporations to fully manage devices, and also to support the management of devices under the BYOD paradigm.

1.5 SECURITY HARDENING AND PREVENTIVE TECHNIQUES

There are several techniques that can be used to increase the security on the mobile device, involving good practices in application development, deployment of security policies at the enterprise level, and usage of existing security features on mobile devices aimed at ensuring its defense against external threats. This section explores some of these techniques.

1.5.1 Application Development Aspects

Developing secure applications is a challenging task, requiring developers to enforce good practices regarding data protection, resource usage, event management, and other aspects that might be exploited or abused by malicious applications, devices, or users. Prevention can also come into play with techniques such as code obfuscation.

1.5.1.1 Data Protection

Usually each mobile operating system offers a few ways of ensuring data protection to some extent, and the same is true for application development. The most common form of securing data is to encrypt it with cypher algorithms. Nevertheless, good encryption may be costly in terms of resources and time, so it might not be suitable for all types of systems and applications. An encrypted device might benefit from the added security, but performance will suffer a slowdown.

Thankfully, most mobile device operating systems offer some alternatives to encryption that applications can use, such as sandboxed environments. We have already mentioned sandboxed environments regarding prevention, and their purpose is, to some extent, to isolate applications from one another and from the system, minimizing the amount of damage that they might cause and the data to which they have access. Conversely, it can also desirable to have some portions of the system that are accessible between applications, where data can be shared.

It should come as no surprise that in light of these considerations, the most common way of securing data is to store it on a mobile device for the least amount of time possible, and later offload it to a cloud service or a secure server that cannot be physically accessed so easily.

In this respect, data can also be protected using the services provided by TEE, where data is stored securely and via trusted hardware functions. Further details can be found at Section 1.5.7.

1.5.1.2 Resource Usage

Mobile devices are often characterized by their lack of a constant power source and more limited computing power than regular computers. This means it is easier to use a significant portion of the system's resources at any given time, and harder to maintain a balanced use throughout the execution time. Furthermore, increased resource usage is directly associated with shorter battery life.

It is important to note that "battery life" has different meanings for rechargeable batteries: it can mean either the length of time a device can run on a fully charged battery or the number of charge/discharge cycles possible before the cells fail to operate satisfactorily. In the previous paragraph we used the former definition.

Moreover, even though the mobile industry has been taking large strides to close the gap between mobile and traditional computers—with some mobile devices boasting as much RAM and CPU capability as found in low to mid-range laptops—the storage paradigm on mobile devices must still rely heavily on remote storage approaches such as the cloud, whereas traditional computers have the luxury of having SSDs and HDDs attached, capable of storing large amounts of data.

1.5.1.3 Event Management

Most mobile operating systems maintain the multitasking aspect that computer operating systems have, although at a more limited level. What this means is that in a mobile device, although several background processes might be running and performing demanding tasks, usually only one application or process can control the foreground. This is not an ironclad rule, as some systems do allow for a couple of applications or processes to coexist in the foreground space. It is, however, a consequence of user interaction with smaller displays and reduced available resources.

This brings us to event management, the response to limited multitasking in mobile devices. Since usually only one application is using the device's foreground space, any other application or service that wishes to interact with the user through the foreground space must announce it's intentions, and this is done with system events. What's more, some systems divulge most of their runtime changes or allow for their tracking through system-wide event broadcasting, which opens the door to smarter event management. An application that requires specific resources can adapt its behavior based on the system's state.

1.5.1.4 Exploits and Malicious Applications, Devices, or Users

As already discussed, application development for mobile systems can require some additional thought beyond writing code, depending on the goals and purpose of the application, on the sensitivity of the managed data, and on the expected resource usage.

If an application that was written disregarding good security practices handles sensitive data, it becomes a potential target of other malicious applications or users, attempting to steal important information. The already mentioned encryption and sandboxing are preventive measures for data protection, generally supported by mobile systems, but they are not infallible. Some forms of malware exploit the host system to attain the encryption keys or to peek and interact with private sandboxed spaces of other applications (or, in some cases, even the system's kernel).

Another way an application can be exploited is by decompiling it and analyzing its source code, as it might reveal clues or vulnerabilities. Code obfuscation is a way of making decompiled code more difficult to interpret by exchanging otherwise helpful method and variable names with nonsensical or misleading alternatives.

1.5.2 Security Policies

It is a common practice for mobile devices and the software running on them to have a list of security recommendations and guidelines. For instance, both Apple and Google have detailed specifications on what can and should be a part of their devices and applications [5,6].

Although there might exist other recommendations to ensure the desired visual design and user interaction, functionality, compatibility, performance, and stability, security policies are important and timeworthy when it comes to striving for a more secure and harder to exploit application.

Let us take into account the following example of the Android security policies:

The application does not dynamically load code from outside the application's Android PacKage (APK).

The application uses strong, platform-provided cryptographic algorithms and does not implement custom algorithms.

The application uses a properly secure random number generator, in particular to initialize cryptographic keys.

Any of these aspects can be exploited by malware or malicious users, so Google has decided to inspect Android applications submitted to the Play Store for potential misuse regarding them, issuing red flags where deemed fit.

One outcome that might arise from the preventive enforcement of security policies is that, since security policies are disclosed on public domain (or at least to registered developers), malicious agents can avoid them upfront, attempting to circumvent them and target other security blindspots or loopholes.

1.5.3 Official Secure Application Stores

Official application markets, often maintained by the companies providing the OS or device ecosystem family, are a secure way of controlling content that runs on mobile devices. Typically, by default the installation of applications from outside these sources is not allowed, for security purposes. On iOS for instance, the applications are subjected to an evaluation process to assess whether they obey the guidelines established by Apple and to detect whether any malicious piece of code is present. This is a technique that reduces the chance of installing malicious applications on the mobile devices by controlling what users can actually have access to.

1.5.4 Operating Systems Recommendations

Entities responsible for operating systems are aware of the problem of application developers who do not follow the best practices in terms of security for a given OS platform. Therefore, on their own initiative they provide development guidelines to guide developers toward adopting adequate practices, in order to eliminate or reduce the application attack surface [20]. The following are some of the most important guidelines on Android are:

Using internal storage. In Android the files created on internal storage are accessible only to the application that created them. To share data with other processes, developers must use content providers, offering read and write permissions to other applications. Furthermore, to provide additional data protection for data that is sensitive, local files may be encrypted.

Permissions. Since Android employs a sandboxing approach, the applications must explicitly share resources and data. In this sense, if they need to access additional features that are not provided by the basic sandbox, they must declare the required permissions. The number of permissions an application requests should be minimized, and special attention must be paid when requesting sensitive permissions. This reduces the risk of permission misuse and makes the application less vulnerable to attackers. Also, it is advisable that the application does not leak any data that it has acquired via a granted permission.

Networking. Networking on Android is similar to other Linux environments, therefore internet application communication should employ HTTPS wherever it is possible, to mitigate the risks of accessing public Wi-Fi networks. Moreover, to ensure that other communications stay secure, SSL/TLS should be employed. Furthermore, the use of existing Android secure tunnel APIs such as `HttpsURLConnection` and `SSLSocket` is encouraged, as opposed to writing custom protocols difficult to verify and maintain over time.

Handling user data. Application developers should minimize the number of APIs that have access to sensitive or personal user data. For instance, if there is no need to store or to transmit user data, it is advisable that the application does not store or transmit it. Instead, the use of hashes is encouraged. This reduces the possibility of exposing data or having attackers attempting to exploit the application.

Interprocess communication. Even though Android is similar to Linux, application developers should use the APIs that Google provides to perform IPC, such as `Intent, Binder, Messenger, Service, BroadcastReceiver`. When the IPC is intended for in-application only use, adequate modifications must be carried out in the application's manifest.

1.5.5 Data Storage

Device encryption is becoming more and more important as users increasingly tend to store personal and private data on their mobile devices. When a device is encrypted, an attacker has no access to the data on the device without the user's PIN or biometric credentials.

Device encryption in Android aims to ensure that even if an unauthorized party has access to the data it will not be able to read it. When a device is encrypted, all the data created by the user is ciphered with symmetric encryption keys prior to being written to disk, and upon a read it is automatically deciphered before being returned to the calling process. There are two methods for device encryption in Android, depending on the OS version.

Since Android 5.0 it is possible to perform full-disk encryption using a single key to protect the whole partition where the user data is. On one hand, this means that the user must provide his/her credentials to access the device's disk, thus making the data on the device secure. On the other hand, this also means that most of the core functionality of the device is not available upon boot, which makes alarms, call reception, and other features unavailable.

Since Android 7.0 it is possible to encrypt the data on the device on a file basis. This allows different files to be encrypted with different keys, meaning they can be unlocked independently. Moreover, devices that support file-based encryption are also able to use a feature called `Direct Boot` that allows encrypted devices to boot to the lock screen directly, with limited resources available. It is then up to application developers to perform the necessary adjustments for providing functionality on the lock screen after a boot on an encrypted device.

As already mentioned, a relevant aspect regarding security hardening strategies for mobile devices has to do with the way information is stored on the devices, as well as managing how or whether other applications or devices may access that information. Encryption of sensitive data before it is stored has already been mentioned, but it can also be stored differently depending on its nature and purpose. For instance, both Android and iOS enable applications to store information in some of the following ways:

Application only. Stored information is only visible to the application that wrote it.

Internal storage file system. Data is stored on the device's internal storage where, depending on the operating system, it might be visible to other applications.

External storage file system. Data resides on a storage medium that can easily be exchanged or removed from the device, and it is visible to all other applications.

Database. All information is stored and managed by a database that, depending on its location, may be visible to other applications. For instance SQlite, the most commonly used mobile system lightweight database, can be set up to encrypt its contents.

Unfortunately, all these storage methods are vulnerable to misuse, and even though the system might ensure that sensitive content is kept private from other applications under normal circumstances, usually the same cannot be assured for rooted devices or hacked user accounts. This means that if an application that handles sensitive data, stored in a secure fashion, is installed on a rooted device or if another application manages to attain root privileges (e.g., privilege escalation) then the application's data might be compromised.

1.5.6 Authentication

Perhaps the most common form of enforcing security on mobile devices is by establishing authentication barriers before the user can access sensitive information or functionalities, of which the most widely available are biometric data (fingerprints, face recognition, etc.), passwords, PIN codes, and patterns. These can also be paired with other forms of authentication such as the device's location (where only certain environments are deemed trustworthy).

There are several types of biometric authentication mechanisms, among which the following are the most commonly cited:

Face recognition. By mapping key points and facial features of an individual's face, a picture or video can be analyzed and the subject identified or verified.

Fingerprint. The most commonly used biometric authentication method, regarded as unique even between twins. Fingerprints do not change over time and can be easily mapped to an identity.

Hand veins. When viewed under infrared light, the veins near a hand's surface become visible, and these can be treated as features, much like fingerprint ridges.

Iris scan. Irises are unique between individuals; however, since they can contract and dilate depending on the ambient lighting, iris scans are often performed under a more controlled environment.

Palm print. Instead of analyzing a digit, the whole palm of the hand is scrutinized, allowing for more features to be identified.

Signature. The user writes his/her signature and its shape and the writing process are analyzed.

Speaker recognition. Unlike voice recognition, which aims to differentiate sounds from words, speaker recognition attempts to differentiate one person from another. This can be achieved by speaking a known word or sentence and analyzing how each individual does it differently.

As one might correctly speculate, not all these authentication methods are equally secure or practical, which is why, in the mobile context, only fingerprint and facial recognition have been adopted by mobile device manufacturers. Moreover, these systems are fallible, with possible false positives and false negatives. As the sensors for these technologies become more compact and reliable, it is only sensible to include them in mobile devices, as they provide a more comfortable authentication experience without compromising security, as opposed to memorizing or carrying around very long and cryptographically secure keys/passwords.

1.5.7 Isolation and Trust

Even though iOS and Android provide application sandboxing capabilities, there are times and contexts that require applications or operations to be executed in further isolation from the system and possibly in a trusted way. "Trusted way ensures that the critical functions of both hardware and software have not been modified. One example could be trust in the input methods, such as keyboard, where it is guaranteed that the inputs provided (e.g., passwords) are not logged, read by malicious or third party applications. The most common approaches to ensure isolation include virtualization and trusted execution environments (TEE).

1.5.7.1 Virtualization

Virtualization aims to decouple the physical hardware from the OS. Inherent in virtualization is the concept of a *hypervisor*, which is a software component responsible for the scheduling of tasks and the management of the device's resources for the virtual machines. There are two categories of hypervisors: a type 1 hypervisor runs directly on the hardware and provides the available resources to the guest OS; a type 2 hypervisor runs on an operating system that provides the I/O resources. Even though type 1 hypervisors offer better performance and profile isolation, they have an increased burden on CPU and memory, since each virtual machine executes its own complete kernel and OS instance. Type 2 hypervisors have a lower degree of isolation between profiles but also offer lower OS switching overhead as certain portions of the host kernel and OS instance are shared. There have been a few attempts to introduce virtualization into mainstream mobile devices but, up to now, this remains a very small niche approach for special-purpose ecosystems.

1.5.7.2 Trusted Execution Environment

The trusted execution environment (TEE) is a secure, integrity-protected processing environment, with memory and storage capabilities [41], that aims to achieve secure computation, privacy, and data protection. The TEE is also associated with other terms, such as secure execution environment (SEE) and dynamic root of trust measurement (DRTM). The SEE aims to guarantee authenticity of code, integrity of runtime states (not tampered), and confidentiality, as data should not be disclosed to unauthorized applications. In this sense the TEE complements the SEE by including the notion of trust, which is also the base of DRTM.

In the TEE there is a trusted platform module (TPM), which allows a system to provide evidence of its integrity and to protect cryptographic keys inside the hardware modules. As TPMs are closely associated with vendors of hardware (e.g., Samsung, LG, Apple), the development of secure applications is not straightforward, since the used TPM APIs are not publicly available to developers. With this in mind, GlobalPlatform [23] specifies trusted end-to-end solutions that facilitate the secure and interoperable deployment and management of embedded applications on secure chips. In particular, the specifications of GlobalPlatform for TEE [24–26] aim to facilitate the development of third-party applications (i.e., not provided by hardware vendors) by addressing trusted client APIs, trusted user interfaces and trusted internal APIs, as illustrated in Figure 1.4.

To use the trusted functionalities provided by TEE, applications need to be split into client application (CA) and trusted application (TA), where the last holds the security-sensitive functions (i.e., implementation of encryption algorithm). As depicted in Figure 1.4, the CA runs in the normal OS (e.g., Android), while the TA runs in the trusted part (i.e., in the specific hardware-supported secure environment) and is responsible for making the interface with the trusted and secure APIs provided by TEE. CA and TA exchange data via

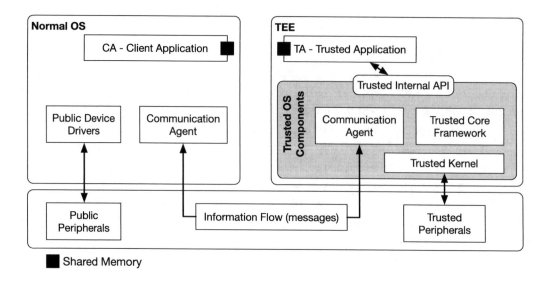

FIGURE 1.4   Trusted execution environment.

shared memory mechanisms. Additionally, the CA implements the necessary functions to initiate a session.

TEE also includes support for securing operations and the retrieval of data from trusted peripherals. Despite the promising features of TEE and a strong investment from vendors, the supported functionalities are still limited at this point. For instance trusted user interfaces only support numerical input from a keyboard in the screen. Additionally, vendors/ manufacturers do not disclose their support of GlobalPlatform specifications, therefore not enabling the development of secure and trusted solutions by application developers.

1.6 MONITORING, DETECTION, AND REACTION

This section focuses on the techniques that can be used to detect malicious events on the mobile device. It will provide an overview of the methods used for data acquisition, also discussing which features must be present on a detection and reaction mechanism.

1.6.1 Monitoring

Monitoring on the device can be performed through the collection of several parameters that provide information regarding the events that are occurring. Such parameters include application permissions [38], system calls, intents, API calls, used devices, and behavioral data such as performed calls, sent messages, taken photos, connected networks, and visited sites, among others.

The monitored parameters are important for performing intrusion detection, as described in Section 1.6.2. Data collection mechanisms may be based on polling mechanisms, where items are collected at certain intervals, or be always active. Mechanisms that are always active constantly listen to system events, for instance by implementing broadcast receivers in Android to collect information from system events (recall Section 1.3.1).

1.6.2 IDS and IDPS (Detection and Reaction)

Intrusion detection systems (IDSs) and intrusion detection and prevention systems (IDPSs) can be used to detect events in a system that suggest that its security has been breached, and also to provide proactive protection mechanisms. They involve the detection, logging, and reporting of the collected information. IDPSs have an extra step that consists in trying to block or stop the detected malicious activity [43]. According to the detection methods employed, they can be categorized as static, dynamic or hybrid:

Static analysis corresponds to the examination of the application's code without executing it, disassembling and decompiling the application and trying to recreate its algorithms through reverse engineering techniques. Analysis of the requested permissions is a common method of static analysis. Nevertheless, this method usually provides low accuracy because application developers often request permissions that are not needed. Moreover, static analysis methods can be fooled by code obfuscation or transformation techniques.

Dynamic analysis usually relies on the collection of data from the devices (CPU usage, network traffic, battery consumption, etc.), which is examined using data mining methods to extract patterns or behaviors that potentially represent normal or abnormal activities [16,17]. These patterns are combined and used to create rules that detect known and unknown malicious activities through similarities with the already known patterns. Dynamic approaches can monitor the behavior of programs during runtime using heuristic-based methods; therefore, the problems related to malicious applications that use obfuscation methods to bypass static analysis are solved. Nevertheless, dynamic analysis can also fail when malicious applications employ evasion techniques.

Hybrid analysis uses methods from both static and dynamic analysis to transcend the drawbacks of each individually. By combining such methods, more robust techniques for malicious application detection are created.

Even though they are important tools for malicious application detection, IDSs have several limitations. For instance, they rely heavily on the information that the tools on the devices provide to operate properly. Their accuracy can be hindered by events on the host such as reboots, modification of system files, and application installation/uninstallation. These events can be detected with success but the nature of the processes which originate them cannot be discovered without additional information on the context. Detection of such events without additional information can easily lead to false positives.

Static and dynamic mechanisms have unique drawbacks that hybrid mechanisms try to improve upon. Nevertheless, hybrid mechanisms may also be thwarted in emulator-based approaches which is a matter of concern.

In static analysis specifically, there can be problems with the detection of leakage of sensitive information when malicious application behaviors do not trigger permission checks or when they use only a single permission to extract sensitive information. When using specification languages, it is important to take into account that the malicious application may have features that are not expressible in the specification language, resulting in detection evasion. Furthermore, there are several technological limitations [45] that are important to address:

Alert generation delays. Even though the agents running on the host generate alerts in real time for the majority of detection techniques, some are only triggered periodically to collect information about events that have already occurred. This means that some events can have a delay between actually occurring and being identified.

Centralized reporting hijacking and delays. Many techniques require the alert information to be forwarded to a centralized management server. It is possible that the communication to a centralized server is intercepted in order to provide false information regarding the malware behavior. Also, by sending the data periodically rather than in real time, it is possible to reduce overhead in the system's components and on the network. Smaller host-based intrusion detection system (HIDS) implementations are more prone to transfer data on a more regular basis, but larger implementations may have larger intervals at which data is transmitted. This can also cause delays in response actions, thus increasing the impact of quickly spreading incidents, such as malware infections.

Host resource usage. These systems require the hosts to have agents running in them to monitor the information, thus needing the host's resources to perform their operations. The agents can have a negative impact on the memory, processor usage, disk storage, network, and file system usage.

Conflicts with existing security controls. Installing an agent on the host can cause conflicts with security controls and other components that also intercept host activity. Therefore, to identify potential conflicts, the agents must be tested on hosts that are running security components such as those present on the hosts on which they will be deployed.

Nowadays, IDPS systems are employed to assure that security is provided in most stringent scenarios. For instance, in mission-critical scenarios like PPDR the usage of such systems complements the other security approaches, such as prevention with dedicated application stores. Indeed, by analyzing the monitored data, IDPS systems are able to provide reaction to threats not known or categorized before [10].

1.7 CONCLUSION

Despite their (often) small form-factor and apparently humble capabilities, (especially when compared with their bigger computing counterparts, such as desktop PCs), mobile device security constitutes a considerable challenge. This is due to factors such as the sheer amount of connectivity alternatives, the dominant usage profiles, or the ease of access or theft, associated with many of the problems that have plagued more traditional computing devices and their software for decades.

We have addressed a plethora of topics in this chapter regarding mobile device security. We started by analyzing the current open challenges such as the evolution of mobile device complexity, and the risks associated with the information generated by the use and ownership of these devices, intending to demystify the common assumption that mobile devices are simple in nature, when in reality they have become more and more complex throughout the years, holding an ever-increasing influence on our daily lives. From there we moved on to the core concepts that should be considered when thinking about mobile device and mobile application security, which can be different depending on the target environment.

Corporate and consumer environments can be very distinct when it comes to safeguarding sensitive information, particularly when considering the implications of enabling compromisable devices in compromisable networks. All of these considerations warrant adequate mobile device management analysis and corresponding policies. This security approach can come in various flavors: from prevention, to monitoring and detection, to reaction. There are of course some entities or platforms already in place for such an endeavor, each with their own strengths and weaknesses, but all with the added benefit of already having proof-of-concept and test user bases to uphold security technologies.

There are also techniques that one can employ without relying completely on a third party's security, among which are some basic considerations regarding application development, the importance of following preestablished security policies and guidelines, how to isolate sensitive data from prying eyes through encryption, and authentication mechanisms, and others.

It is our hope that, after reading this chapter, the reader has gained more insight into the sheer number of possible attack vectors, with more constantly being added to the list, as well as the means with which to hinder them during application development or mobile device management.

REFERENCES

1. A. Bello Garba, J. Armarego, D. Murray, and W. Kenworthy. Review of the information security and privacy challenges in bring your own device (BYOD) environments. *Journal of Information Privacy and Security*, 11(1):38–54, 2015.
2. M. Alsaleh, N. Alomar, and A. Alarifi. Smartphone users: Understanding how security mechanisms are perceived and new persuasive methods. *PLOS ONE*, 12(3):1–35, 2017.
3. Android Developers. Permission Element, 2018. Available at: https://developer.android.com/guide/topics/manifest/permission-element (last accessed September 17, 2018).
4. Apple. Apple business, 2017. Available at: https://www.apple.com/lae/business/ (accessed September 17, 2018).
5. Apple. iOS application security guidelines, 2018. Available at: https://www.apple.com/business/site/docs/iOS_Security_Guide.pdf (accessed September 17, 2018).
6. Google Android Application Quality Guidelines. Core app quality, 2017. Available at: https://developer.android.com/docs/quality-guidelines/core-app-quality (accessed September 17, 2018).
7. Arm. Security on Arm Trustzone, 2018. Available at: https://developer.arm.com/technologies/trustzone (last accessed September 17, 2018).
8. M. Becher and F. C. Freiling. Towards dynamic malware analysis to increase mobile device security. In *Proc. of SICHERHEIT, P-128*, pages 423–433, 2008.
9. Google Security Blog. From chrysaor to lipizzan: Blocking a new targeted spyware family, 2017. Available at: https://security.googleblog.com/2017/07/from-chrysaor-to-lipizzan-blocking-new.html (accessed September 17, 2018).
10. P. Borges, B. Sousa, L. Ferreira, F. Sagheczchi, G. Mantas, J. Carlos Ribeiro, L. Cordeiro, and P. Simoes. Towards a hybrid intrusion detection system for android-based PPDR terminals. In *3rd Workshop on Security for Emerging Distributed Network Technologies, DISSECT 2017, IM*, May 2017.
11. A. Bose, X. Hu, K. G. Shin, and T. Park. Behavioral detection of malware on mobile handsets. In *Proceedings of the 6th International Conference on Mobile Systems, Applications, and Services, MobiSys '08*, pages 225–238. ACM, New York, NY, USA, 2008.
12. J. R. Boyd. Presentation: The essence of winning and losing, 1995, pp 3. Available at: http://pogoarchives.org/m/dni/john_boyd_compendium/essence_of_winning_losing.pdf (accessed September 17, 2018).
13. B. Bulgurcu, H. Cavusoglu, and I. Benbasat. Information security policy compliance: An empirical study of rationality-based beliefs and information security awareness. *Management Information Systems Quarterly*, 34(3):523–548, September 2010.
14. I. Burguera, U. Zurutuza, and S. Nadjm-Tehrani. Crowdroid: Behavior-based malware detection system for android. In *Proceedings of the 1st ACM Workshop on Security and Privacy in Smartphones and Mobile Devices - SPSM '11*, page 15, 2011.

15. C. Willemsm, T. Holz, and F. Freiling. Toward automated dynamic malware analysis using cwsandbox. *IEEE Security and Privacy*, 5(2):32–39, 2007.

16. V. Chandola, A. Banerjee, and V. Kumar. Anomaly detection. *ACM Computing Surveys*, 41(3):1–58, 2009.

17. J. Cheng, S. H. Y. Wong, H. Yang, and S. Lu. SmartSiren: Virus detection and alert for smartphones. In *Proceedings of the 5th International Conference on Mobile Systems, Applications and Services – MobiSys '07*, page 258, 2007.

18. CloudFlare. Mobile ad networks as ddos vectors: A case study, 2015. Available at: https://blog.cloudflare.com/mobile-ad-networks-as-ddos-vectors/ (accessed September 17, 2018).

19. J. Concepcion, J. Chua, and G. Siy. Securing android byod (bring your own device) with network access control (nac) and mdm (mobile device management). In *DLSU Research Congress 2015*, De La Salle University, Manila, Philippines, 2015.

20. Android Developers. App security best practices, 2018. Available at: https://developer.android.com/topic/security/best-practices (accessed September 17, 2018).

21. C. J. D'Orazio, K. K. R. Choo, and L. T. Yang. Data exfiltration from internet of things devices: iOS devices as case studies. *IEEE Internet of Things Journal*, 4(2):524–535, 2017.

22. A. M. French, C. Guo, and J. P. Shim. Current status, issues, and future of bring your own device (BYOD). *Communications of the Association for Information Systems*, 35(10):191–197, 2014.

23. GlobalPlatform. Available at: https://globalplatform.org/ (accessed September 17, 2018).

24. GlobalPlatform. Globalplatform device technology TEE client API specification table of contents, 2010. Available at: https://globalplatform.org/specs-library/tee-client-api-specification/ (accessed September 17, 2018).

25. GlobalPlatform. Globalplatform device technology TEE internal API specification, 2011. Available at: https://globalplatform.org/specs-library/tee-internal-core-api-specification-v1-1-2/ (accessed September 17, 2018).

26. GlobalPlatform. Trusted user interface API v1.0, 2013. Available at: https://globalplatform.org/specs-library/trusted-user-interface-api-v1/ (accessed September 17, 2018).

27. Nico Golde. SMS Vulnerability Analysis on Feature Phones. *PhD thesis*, Technical University of Berlin, Berlin. Germany, January 2011.

28. Google. Android Enterprise, 2018. Available at: https://www.android.com/enterprise/ (accessed September 17, 2018).

29. G. Greene and J. D'Arcy. Assessing the impact of security culture and the employee-organization relationship on is security compliance. In *5th Annual Symposium on Information Assurance (ASIA'10)*, page 1, 2010.

30. M. Halilovic and A. Subasi. Intrusion detection on smartphones. *arXiv e-print 1211.6610*, 2012. Available at: https://arxiv.org/ftp/arxiv/papers/1211/1211.6610.pdf (accessed September 17, 2018).

31. J. T. Jackson and S. Creese. Virus propagation in heterogeneous Bluetooth networks with human behaviors. *IEEE Transactions on Dependable and Secure Computing*, 9(6):930–943, 2012.

32. C. M. Jones. *Utilizing the Technology Acceptance Model to Assess Employee Adoption of Information Systems Security Measures*. Nova Southeastern University, Fort Lauderdale, FL, USA, 2009.

33. H. Marques et al. Wireless public safety networks 1: Overview and challenges. In *Next-Generation Communication Systems for PPDR: The SALUS Perspective*, pages 49–93. Elsevier, New York, NY, USA, 2015.

34. S. Mazumdar and A. Paturi. Tamper-resistant database logging on mobile devices, 2011. *World Congress on Internet Security (WorldCIS-2011)*, London, 2011, pp. 165–170.

35. Microsoft Buxton collection. Simon Cellular Phone/PDA (Personnal Digital Assistant), 1994. Available at: https://www.microsoft.com/buxtoncollection/detail.aspx?id=40 (accessed September 17, 2018).

36. A. Mylonas, A. Kastania, and D. Gritzalis. Delegate the smartphone user? Security awareness in smartphone platforms. *Computers and Security*, 34:47–66, 2013.

37. L. Constantin, PC World "New Windows malware tries to infect Android devices connected to PCs", 2014. Available: https://www.pcworld.com/article/2090940/new-windows-malware-tries-to-infect-android-devices-connected-to-pcs.html (accessed September 17, 2018).

38. Android Open Source Project. Runtime Permissions, 2017. Available at: https://source.android.com/devices/tech/config/runtime_perms (accessed September 17, 2018).

39. Radware. Mobile security threats on the rise as hackers can launch ddos attacks on their mobile phones, 2016. Available at: https://security.radware.com/ddos-threats-attacks/cyber-attacks-in-the-palm-of-your-hand/ (accessed September 17, 2018).

40. Dimensional Research. The growing threat of mobile device security breaches a global survey of security professionals, 2017.

41. M. Sabt, M. Achemlal, and A. Bouabdallah. Trusted execution environment: What it is, and what it is not. In *IEEE Trustcom/BigDataSE/ISPA*, Aalto University, Helsinki, Finland, volume 1, pages 57–64, 2015.

42. Samsung. Whitepaper: Samsung Knox - Security Solution, May 2017. Available at: https://www.samsungknox.com/docs/SamsungKnoxSecuritySolution.pdf (accessed September 17, 2018).

43. K. Scarfone and P. Mell. *Guide to Intrusion Detection and Prevention Systems (IDPS) Recommendations of the National Institute of Standards and Technology*. Technical Report, NIST, Gaithersburg, MD, USA, 2007.

44. S. Smalley and R. Craig. Security enhanced (SE) android: Bringing flexible MAC to android. *NDSS Symposium*, 2013. Available at: https://www.ndss-symposium.org/ndss2013/ndss-2013-programme/security-enhanced-se-android-bringing-flexible-mac-android/ (accessed September 17, 2018).

45. G. Stoneburner, A. Goguen, and A. Feringa. NIST special publication 800-94 revision 1. *Risk Management Guide for Information Security*, page 95, September 2012.

46. Teclib. Flyve mdm, open source mobile device management solution, 2017. Available at: https://flyve-mdm.com/ (accessed September 17, 2018).

47. J. Vila and R. J. Rodríguez. *Practical Experiences on NFC Relay Attacks with Android*, pages 87–103. Springer International Publishing, Cham, Switzerland, 2015.

CHAPTER 2

Model-Driven Design
of Connectivity-Aware
Mobile Applications

Steffen Vaupel, Gabriele Taentzer, and Michael Guckert

CONTENTS

2.1 INTRODUCTION

Reliable operation of mobile applications during movement in space is a challenge for mobile application developers [44]. When a mobile device switches from one radio cell to another [55], loss or limitation of connectivity is not unusual and should be handled by the architecture of mobile applications. Mobile application developers rarely implement online- and offline-capable mobile applications, although this would considerably increase the usability of such applications and users are increasingly interested in online- and offline-capable mobile applications [18]. Besides dealing with interrupted network access, online- and offline-capable mobile applications are also more robust against server failures and can reduce data traffic by using replication and caching of data objects.

Constructing mobile applications that are both online- and offline-capable requires additional architectural components for replication, synchronization, reintegration, and local transaction management. If data can be modified locally, mobile applications can avoid or prevent the execution of potentially conflicting transactions [39], and thus ensure conflict-free synchronization/reintegration of modified data objects.

Mobile application developers have to acquire considerable knowledge about online- and offline-capable transaction management before they can design a mobile application that possesses online- and offline-capabilities. During the initiation phase of a mobile application development project, software development teams often have to chose an appropriate architecture and a software platform (e.g., heading either toward a native and standalone mobile application architecture in Android/iOS or a platform-independent web-based mobile application) [29,60]. This choice thereafter dominates the development process and may cause limitations. Upon being forced to decide whether to implement an online-only or offline-only mobile application variant, mobile application developers must thoroughly understand the relevant use cases, for example, processes and accessed data objects of the planned mobile application. Currently, there is no conceptual tooling for evaluating mobile application designs regarding their online and offline capability. Mobile application developers do not get any support in assessing whether the effort to implement such an online- and offline-capable mobile application would be reasonable; for example, there is no assistance to predict whether the throughput would increase considerably compared

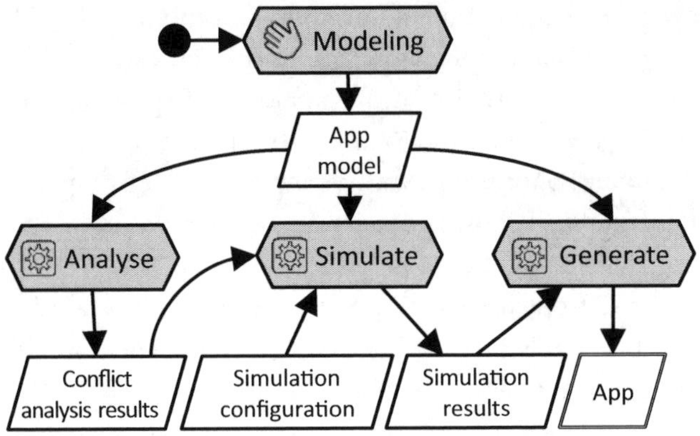

FIGURE 2.1   Model-driven design process of online- and offline-capable mobile applications.

with an online-only or offline-only mobile application variant. Furthermore, the improper utilization of an online- and offline-capable architecture can deteriorate the throughput because data objects might have to be locked by mobile application users being offline. Besides, if a mobile application developer decides to realize an online- and offline-capable mobile application, further issues that may emerge include differing support of consistency and conflict levels between the mobile application and the transaction system. Accepting lower consistency of data objects increases the throughput of transactions, but it may also lead to conflicts during synchronization/reintegration. Mobile application developers are often ambivalent about which conflict and consistency level should be favored as throughput and further implications are quite hard to predict.

This chapter aims at a process that respects three aspects (cf. Figure 2.1): first, we propose to model the core data structures, the behavior of the mobile application, and the graphical user interface (GUI) by using a domain-specific modeling language (*modeling step*). We name such a model *app model*. Based on this concise formulation of mobile application data, behavior, and GUI, a static conflict analysis can be conducted (*analysis step*). This analysis identifies all potential conflicting processes that must be managed in a connectivity-aware way. The model analysis shows the overall capability of a mobile application (i.e., its app model) to be operated in different connection states (i.e., online and offline).

Second, potentially conflicting processes may be tolerable in certain scenarios—for example, a relatively low number of mobile users act on a huge set of data objects and conflicts will probably occur only rarely and can thus be tolerated. By using the results of the static conflict analysis and a simulation configuration, a model-based simulation system (*simulate step*) can predict the number of actually conflicting transactions* for a custom app model. Developers can test different (expected) operational conditions without having to implement the planned mobile application.

Third, the design process allows the generation (*generate step*) of prototypical online- and offline-capable mobile applications. The goal of this software prototype generation is to provide an evolutionary prototype [32] that can be studied and extended manually. The

---

* We will name instantiated, that is executed processes, *transactions*.

generated prototype follows the notion of horizontal prototyping [25]; that is, specific layers of online- and offline-capable mobile applications are built for reuse. More precisely, the generated prototype focuses on architectural components (e.g., application logic, replication, and synchronization/reintegration functionality) that are required [68] to implement online- and offline-capable mobile applications, with a simple generated graphical user interface that may be replaced in further iterations. The generation step builds upon an existing framework for the model-driven development of mobile applications [73] (cf. [71,74], and [72]).

This chapter is structured as follows. We present the foundation of connectivity-aware mobile applications in Section 2.2. Next, we present the domain-specific modeling language applied here and the model elements for modeling connectivity-awareness to be used (Section 2.3). In Section 2.4, we explain how model analysis, design evaluation, and prototype generation works in more detail. Finally, Section 2.5 presents related work and concludes this chapter.

2.2 SOFTWARE ARCHITECTURES OF CONNECTIVITY-AWARE MOBILE APPLICATIONS

Creating online- and offline-capable mobile applications in a traditional way requires a mobile application developer to possess extensive knowledge of software architectures of connectivity-aware mobile applications. Although most architectural aspects are set implicitly by our model-driven development infrastructure, that is, the code generator, we will present these features beforehand so that the reader is familiar with the most suitable software architectures for realizing online- and offline-capable mobile applications.

2.2.1 Classification of Mobile Applications

Since mobile applications are used in a broad range of application areas, they have different requirements with respect to offline and online capability. During requirement analysis of a new mobile application development project, mobile application developers should reflect about requirements regarding the different connection states of the mobile application. These requirements determine a particular architectural design (cf. Book et al. [28]) that contains different architectural features.

A high-level classification of mobile *application types* might distinguish between *standalone systems*, *information systems*, *transaction systems*, and *communication systems* (see Figure 2.2). A *standalone system*, for example, a calculator application or gaming application, usually requires no data objects from other systems. An *information system*

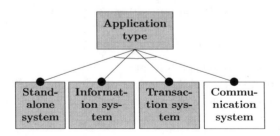

FIGURE 2.2   Manifestations of the feature *Application type.*

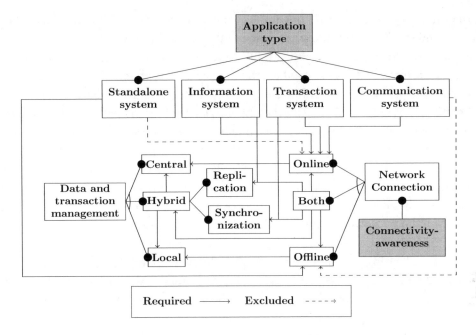

FIGURE 2.3 Architectural implications of the *Application type* and the required degree of *Connectivity-awareness*.

implies that a mobile application, that is, a mobile application user, reads remote data objects (e.g., passenger information system, dictionary, and encyclopedia) in a synchronous or asynchronous way from a local or central storage, while a *transaction system* is reading and writing data objects from/to a back-end server. Transaction systems usually involve more than one mobile client, that is, they are multiuser transaction systems. Finally, *communication systems* also receive and send data like transaction systems but require additional real-time service quality. The different manifestations of the *application type* feature depend on as well as influence other architectural features. Figure 2.3 shows required/excluded features of the network connection (e.g., *online, offline, both*) for different types of mobile applications. Moreover, these connection features imply different features (e.g., *replication, synchronization/reintegration*) of the *data and transaction management* of mobile applications. For example, a *standalone system* does not require any online connection but needs some kind of *local* data and transaction management. In contrast, a *communication system* requires an always-*online* network connection and excludes *offline* operations. The data and transaction management of a communication system is provided remotely by a *central* service (e.g., a back-end server).

2.2.1.1 Application Domain and Example Applications

To illustrate the application of online- and offline-capable mobile applications, we present example domains in which such mobile applications can be used profitably, followed by two application scenarios. These application scenarios are used as the running examples in this chapter.

The three biggest retailers [8] in Germany in 2017—Amazon [9], Otto [14] and Zalando [16]—provide mobile applications for different platforms. However, none of them supports transactions for viewing or ordering products and mobile payment in a disconnected mode.

Offering a replicated digital product catalog has a high potential. Users may view and order products while they are offline and can send orders later to the back-end server when the connection is reestablished. An inherent problem within this setting is the limitation of mobile transaction methods for offline payments [65]. Since an order often requires payment in advance, secure payment is a crucial component of most e-commerce applications (e.g., 80% of popular online shops provide PayPal [22] as a payment service [8]). Payment transactions usually contain an online clearing, and such transactions are not allowed to be performed offline. Thus, mobile e-commerce applications for offline usage require offline payment methods. However, involved data objects are very generic (e.g., mostly reflecting some kind of account object) and operation on such data objects are rather simple.

In industrial settings, the problems are slightly different: mobile applications often support or substitute for manual tasks such as keeping inventory, order picking, or maintenance logging. Data objects involved are tailored to the surrounding business processes or real-world objects. Likewise, operations on such data objects are often more complex than in e-commerce and payment scenarios.

To sum up, mobile applications deal with either aggregated values (i.e., account balance and warehouse stock) having a set of rather simple operations (e.g., increment and decrement) or custom objects (as they occur in, for instance, a car rental) having a set of complex operations (e.g., pick-up, refuel, and event of damage). A data object containing only an aggregated value attribute is called a *summable object*; otherwise, it is to be called a *custom object*. If the set of operations contains at least one operation that alters data objects, there is potential for conflict, because data objects are accessed concurrently in multiuser environments. Online- and offline-capable architectures are required to deal with this conflict potential, in particular with the conflict-free management of such transactions across connection contexts.

In the following, we will take a closer look at two specific applications to demonstrate the different ways in which data objects can be accessed: a payment application (covering an example for a summable object with repeatable operations) and a course booking application for fitness studio members (dealing with custom objects and more complex operations). We start the discussion of each application with the assumption that there is no replication for offline usage. This corresponds to a traditional client–server architecture.

2.2.1.1.1 Payment App   First, we consider applications for making mobile payments such as Apple Pay [19], Google Wallet [21], or PayPal [22]. Traditionally, a banking account is administered on a server at the banking site. The dataset is mainly a single value (aggregate), namely the account balance. This singleton is a so-called hot spot because every transaction changes or accesses this value. The set of transactions in a banking account is very limited (i.e., cash withdrawal, cash deposit, debit, credit, and getting account balance). For each banking account, the number of users is also quite limited. The users of a banking account are the bank itself and at least one bank client. The bank itself arranges debit and credit payments from internal to internal and external accounts. Banking clients may perform cash withdrawals or deposits. We assume that the account is a credit account which may never get into debit state. Transactions causing a negative balance provoke a conflict situation. A mobile variant of such a (simplified) mobile payment application shall be ready for operation even if the network connection is interrupted, that is, if the server with the account object is not reachable. The aforementioned payment applications are currently unable to operate in a connectivity-aware manner. The following example demonstrates a money transfer transaction and the impact that a loss of network connectivity has in an online-only variant of the mobile application:

Example (Money Transfer)

We consider the following use case: a banking client wants to transfer money to another banking client from her mobile phone by using near-field communication (NFC). Most payment applications support this use case if both banking clients are online. The transaction is carried out as follows: The creditor delivers her account information to the debtor via NFC. The debtor sends a corresponding payment order to the back-end server as online transaction. The bank checks the cover of the payment order and executes the transaction. Finally, the creditor updates the account balance (again as online transaction) and confirms the transaction. If one of the mobile clients happens to be offline, the payment cannot be executed.

2.2.1.1.2 Course Booking App   As the next example, we consider a course booking application for fitness studio members. Examples of such applications are GymSync [12], GymJam [11], and BookFit [10]. Registered fitness studio members can select course spots to practice particular exercises—such as yoga and Pilates—within certain time slots. The set of operations is very limited but, compared with payment applications, very specific (e.g., making or canceling a reservation or checking if a course spot is available). Conflicts are very likely because each member may select every course spot. Studio members may set course preferences to indicate which courses they will select in the future. We assume that the course booking application never allows overbooking of a course spot (conflict situation). The aforementioned mobile applications are presently not connectivity-aware. Hence, we are interested in an online- and offline-capable variant of such a (simplified) course booking application. The following example shows a course reservation transaction and the impact that a loss of network connectivity has in an online-only variant of the mobile application:

Example (Reservation of a Course)

If a fitness studio member is online and a selected course spot is available, a reservation transaction can be processed. No transaction can be executed if the mobile client is offline.

2.2.2 Mobile Transaction Models

Transaction models describe how a transaction manager (TM) handles transactions that are asked to be performed on the database (DB). For example, a transaction model might require concurrently executed transactions to have *equal* effects as the sequential execution of the same transactions, that is, they must be serializable. The serializability of simultaneously executed transactions can be ensured by obtaining different correctness criteria that are based on read/write conflict definitions [27]. The transaction manager (more precisely, a scheduler as part of it) can check the serializability of the scheduled transactions dynamically during the execution of concurrent transactions. If a schedule of simultaneously executed transactions is not equivalent to a sequential execution, some of the transactions have to be canceled.

However, such a transaction model is not applicable to the mobile applications domain because different mobile user transactions run on different mobile devices. Moreover due to their interrupted network connection, they cannot send any information about the accessed data objects. Thus, any kind of anomaly, for example reading of inconsistent values, can occur while using a traditional standard transaction model in the domain of mobile applications.

Mobile transaction models deal with this problem and provide further concepts to cope with locally performed transactions of different mobile users on isolated copies. Such models implement three abstract functions: (i) First, a mobile transaction model describes how data objects are replicated from a primary data storage (*Replication*). This is a prerequisite for operating offline. (ii) Second, a mobile transaction model defines which transactions are allowed on replicated copies (*Operation*) to ensure different properties (e.g., consistent reading of data objects, conflict-free synchronization/reintegration). (iii) Third, a mobile transaction model defines how locally modified copies can be synchronized/reintegrated to the set of primary copies (*Synchronization/Reintegration*).

Hirsch et al. [40] survey several mobile transaction models and compare them on the basis of typical requirements for this application domain ([35,47,68,]). Serrano et al. [63,64] and Panda et al. [54] analyze the existing approaches with respect to the well-known ACID (atomicity, consistency, isolation, durability) paradigm. Mutschler and Specht [51] divide mobile transaction models either into *first-class transaction models* that process transactions offline but still need to be online to commit the transaction, or *second-class transaction models* that process transactions offline.

The state of the art provides many approaches for conflict-allowing and semi-offline operating mobile transaction models such as the *Kangaroo transaction model* [35], the *Preserialization transaction management technique (PSTMT)* [34], the *Prewrite transaction model* [50], the *Two-tier transaction model* [39], the *Clustering transaction model* [56–58], the *Reporting and co-transactional model* [31], and the *Isolation-only transaction model* [49]. For this reason, we will focus more on mobile transaction models that are conflict-free and able to finish a local transaction while being completely offline.

The two following examples show that conflict-allowing mobile transaction models are inappropriate in our scenarios:

Example (Payment App)

A debit transaction decreases the replicated *account* value of the debtor and increases the replicated *account* value of the creditor. The application locally checks the coverage of the replicated *account* value. The transaction can happen offline via NFC. Later on, the banking clients (debtor and creditor) reprocess the debit or credit transaction on the primary copy to synchronize the account balance (online transaction). However, if the debtor withdraws money and changes the primary copy before executing the reintegration, the coverage of the account cannot be ensured as the bank is unaware that the customer has already transferred money from the replicated *account*. Since the account may be in the debit state, a conflict may arise.

Example (Course Booking App)

The fitness studio member uses a copy of the entire data set, that is a copy of all course spots. A reservation transaction checks whether a course spot would be unselected by other members and selects it. Later on, the fitness studio member(s) synchronize changed course spots with the primary copies. If another fitness studio member selects the same course spot, the transaction of one member gets lost during synchronization. Nevertheless, both members get a local commitment of their transaction. Since a course spot may be overbooked, a conflict may arise.

As seen in both examples, conflicts may occur if a single shared data object (Payment app) or a shared set data objects (Course booking app) is replicated in a trivial way, that is, if the one-to-one copy and conflicting processes are not limited in any way. Hence, the main idea of conflict-free mobile transaction models is to split the concurrently accessed data objects in such a way that local modification cannot lead to any conflict during the synchronization/reintegration. As we will see later (Section 2.4.1), the deactivation of potentially conflicting processes can also guarantee a conflict-free operation. We have selected the *Keypool transaction model* and the *Escrow transaction model* for a more detailed description.

2.2.2.1 Keypool Transaction Model

Replication: The *Keypool* transaction model [2,4] is well suited to accessing and modifying custom objects. The basic idea of the *Keypool* transaction model is to split the entire dataset of custom objects into subsets that are distributed among participating mobile clients. Here transactions are not restricted when being performed offline, but the set of available data objects is limited. Figure 2.4 illustrates a split of data objects to three mobile clients within the replication step. Every client gets a subset of data objects that is exclusively replicated.

Operation: When a mobile client is offline, all processes of a mobile application are active, but they are not applicable to all data objects due to the client-specific allocation of data objects.

Synchronization/Reintegration: Within the synchronization step, data objects are reintegrated into the primary copies through *image-based* synchronization. Independently of the transactions that have been performed offline, the result can be adopted by substituting the value of the primary copy for the value of the changed replica (i.e., the *image*).

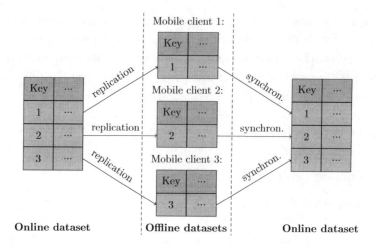

FIGURE 2.4   Keypool replication and synchronization.

Example (Course Booking App)

Figure 2.5 shows the data model of the course booking app. It contains course spots and persons. Course spots are custom objects. This kind of modeling allows using the Keypool approach. However, it may lead to a large set of data objects (e.g., course spots) that is difficult to handle. The reservation of a course spot is made by setting the participant pointer to a person. Owing to the exclusive allocation of custom objects to mobile clients, conflicts cannot occur. However, if the requested object (e.g., course spot) is not allocated to a mobile user's device in a disconnected state, the mobile user does not have any benefit from the used mobile transaction model. The transaction cannot be carried out locally.

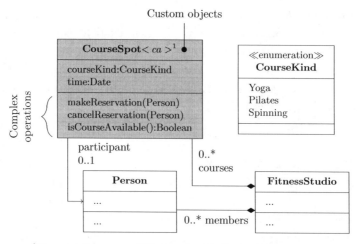

[1]The annotation $< ca >$ will be introduced in Section 2.3.5.3.

FIGURE 2.5   Data model of the course booking app.

2.2.2.2 Escrow Transaction Model

Replication: The *Escrow* transaction model [46,52] is well suited to accessing and modifying aggregate data. The basic idea of this method is to restrict transactions, especially in the domain of their arguments when being performed offline. While the Keypool transaction model splits the dataset into subsets and consequently may provide empty dataset or datasets that are too small, the Escrow approach always provides the full dataset, but splits the values of the data objects. Figure 2.6 shows the replication scheme of data for two mobile clients. The values 90/30 are divided in such a way that every mobile client receives 45/15. Every mobile client gets a full as well as a modified copy of the dataset.

Operation: When a mobile client is offline, all processes of a mobile application are active, but some arguments (e.g., debit an amount of 90) might fail due to the modified copies.

Synchronization/Reintegration: Since every object is transformed at the replication step, conflicts cannot occur at the reintegration step. Considering mobile payment, for example, the debit transaction may cause conflicts. Therefore, the domain of its argument is restricted in such a way that only small amounts may be withdrawn. One possible strategy is to equally distribute the amount among all participating mobile clients, as shown in the aggregate values in Figure 2.6. Since several mobile clients may change the same value, this strategy always guarantees conflict-free reintegration afterwards.

An *image-based* integration does not work here as either one or another image can be written back to the primary copy, but not both. The other values would be lost (called lost update [24,26]). Thus, the reintegration of changed values should be based on a *transaction-based* approach. It collects all transactions performed offline and replays them on the primary copy. The repeated transactions must have the same effects as being performed offline, but they usually do not achieve the same value on the primary copy as in the replicated copies. This property is called *semantical serializability* [37,53]. To ensure it, all operations must be repeatable (such as decrement and increment) and their semantics of restricted values must correspond to the one of the nonrestricted ones. The PRO-MOTION

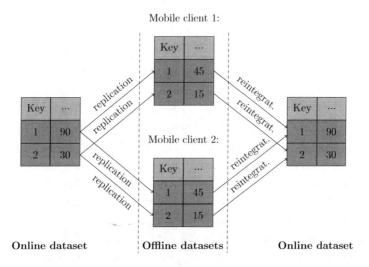

FIGURE 2.6 Escrow replication and reintegration.

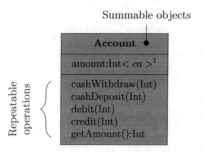

[1]The annotation $< ca >$ will be introduced in Section 2.3.5.3.

FIGURE 2.7   Data model of the payment app.

transaction model by Walborn and Chrysanthis [76] is an extension of the Escrow method and employs so-called *compacts*, which encapsulates access methods, state information and consistency constraints, to allow local conflict-free transactions.

Example (Payment App)

Figure 2.7 shows the data model of the payment app. It consists of the class *Account* only. A valid instance of this data model may have only one *Account* object. Thus, mobile clients may share this object, particularly the attribute *amount*.

2.2.3 Generic Extended Client–Server Model

After discussing different mobile transaction models, mobile application developers are interested in how these mobile transaction models can be applied to a concrete architectural design. As seen from the Keypool and Escrow transaction models, each of these models uses its own replication and synchronization/reintegration method. The authors of the various mobile transaction models have originally intended every mobile transaction model to be used independently. However, each implementation of an architecture that uses a mobile transaction model follows an architectural model called the *extended client–server model* (cf. Satyamaran [61] and Pitoura and Samaras [59]). Such an extended client–server model contains components for *replication* and *synchronization/reintegration* as well as a local database on the client side—this is required by all mobile transaction models. The usual proceeding during the development of an online- and offline-capable mobile application follows a manual implementation of the extended client–server model which, in turn, implements only one mobile transaction model at the same time.

Since we are heading toward a model-driven development approach that can provide different mobile transaction models (at the same time), a first step in this direction is the creation of a generic version of the extended client–server model, as described in our earlier work [75]. In the following, we will describe the main architectural components and the working model of this *generic extended client–server model*.

Figure 2.8 shows the components of the generic extended client–server model. The *application logic* of the client is extended by a *local transaction manager (TM)* that delegates all transactions either to the local database management system (DBMS) or to the central

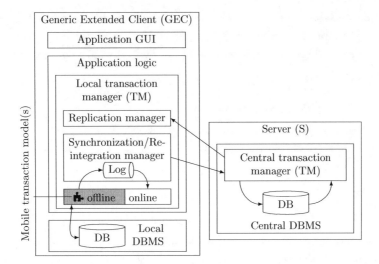

FIGURE 2.8   Generic extended client–server model. DB, database; DBMS, database management system.

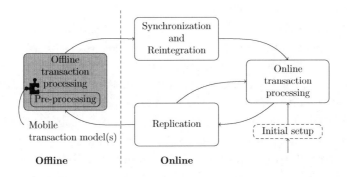

FIGURE 2.9   Working model of the local transaction manager.

one in the server, depending on its connection state. Moreover, the *local transaction manager* comprises a *replication manager*, which is responsible for data replication to be used while mobile clients are offline. Transactions processed offline are logged by the *synchronization/reintegration manager*. Later, this *log* is used to reprocess the transactions performed offline.

Figure 2.9 shows the working model of the local transaction manager being used in our architecture. *Replication*, *synchronization* and *reintegration* are independent generic steps, while *offline transaction processing* implements the mobile transaction model including some preprocessing (gray-colored area). Different mobile transaction models, such as Keypool and Escrow, may be plugged in and work independently of the steps performed online.

2.2.3.1 Data Modeling and Initial Setup

As shown later, data modeling of our model-driven development approach will follow the notion of object-oriented data modeling. Hence, if a database is used underneath (at the client- and server-side), a class model can be translated into a relational data model by object-relational mappers (ORMs). ORM frameworks can create empty database schemes from class models and convert objects into table rows. Similarly, database records may be translated back into object structures.

2.2.3.2 Online Transaction Processing

The online transaction processing step of our generic architecture works like an online-only client–server architecture; that is, a standard transaction model is used for the execution of online transactions.

2.2.3.3 Replication

The generic version of the extended client–server model will not follow the particular replication method of a mobile transaction model but will copy the entire set of data objects to the mobile clients (*full replication*). Furthermore, data objects will not be modified during the replication step. The replication step is triggered by the clients. This is called *pull-based replication*, which is opposite to *push-based replication* [41].

2.2.3.4 Offline Transaction Processing

Offline transaction processing is the crucial part of our generic architecture. Replication is required to operate offline, but it also involves the risk of synchronization/reintegration conflicts. The selected mobile transaction models should ensure conflict-free synchronization/reintegration and commit immediate transactions without having to wait for reconnection. As seen before, the key to conflict-free mobile transaction models is clever distribution of custom objects (Keypool) or splitting aggregates from summable objects (Escrow). However, the full replication of our generic extended client–server model will be contradictory to this conflict-avoiding replication. Thus, depending on whether the modified data objects (i.e., their classes) are marked as connectivity-aware (<ca>) or not, a local preprocessing method will eliminate the custom objects or part of aggregates that would not be available if the mentioned mobile transaction models are used purely without any combination. Based on this preprocessing method, all processes of the mobile application can be executed and no conflicts will occur during the execution of the synchronization/reintegration step.

We will emphasize that conflict-free offline transaction processing relies on a client-specific separation of data objects—as shown in Figures 2.4 and 2.6—to avoid conflicts. A quite contrary approach is based on the isolation of potentially conflicting processes that are executed on the current accessed data objects. Since the behavior of the mobile application could not currently be modeled due to the absence of a suitable domain-specific modeling language, an analysis of conflicts and an isolation of conflicting processes are not feasible. We will later provide the necessary conditions, including the fact that our generic extended client–server model provides both (i) conflict-free operations based on the *isolation of concurrent accessed data* and (ii) the *isolation of potentially conflicting processes*.

2.2.3.5 Synchronization and Reintegration

Modified replicas must be synchronized with the primary copy and the replicas of the other mobile clients. The generic extended client–server model proposed uses a hub-oriented synchronization/reintegration scheme. Every mobile client synchronizes the modified replicas with the primary copy in the server (hub). Since the server does not push announced changes to the mobile clients, mobile clients are asked to pull all changes from the server.

Image-based synchronization: Most of the commercial products use an image-based synchronization which changes the primary copy with the image (or value) of the modified replica. It is not necessary to know the semantics of the executed processes for this synchronization method. Unfortunately, the image-based synchronization requires an $(n - 1)$ consistency, which means that only one replica may have been changed. More than one modified copy (consistency less than $[n - 1]$) cannot be synchronized with the primary copy (i.e., lost updates would occur). Image-based synchronization is sufficient for the Keypool method because an object is modified by at most one mobile client.

Transaction-based Reintegration: This is a rarely used reintegration method that synchronizes the primary copy by reprocessing all the transactions performed on the replicas. A major advantage of the transaction-based reintegration is that no explicit consistency level is required. A set of fully inconsistent copies can be synchronized with the primary copy by reprocessing the transactions that have been carried out locally. But some practical problems inhibit the application of transaction-based reintegration—while the Escrow transaction model requires only simple transactions (e.g., increment/decrement), data-driven mobile applications usually perform complex transactions on local data replicas. To use transaction-based reintegration, these transactions must explicitly be known so that they can be sent to the server for reprocessing. Current commercial solutions of mobile database management systems (mDBMS) are strictly separated from the application logic and thus the semantic of the transactions is unknown.

2.2.3.6 Commercial Products

The number of mDBMS products is still limited, outdated, and very homogeneous in terms of used mobile transaction models. OracleLite [6], IBM DB2 Everyplace [1], Microsoft SQL Server CE [13], and Sybase Adaptive Server Anywhere [3] are some of these products. Their architecture generally corresponds to the *extended client–server model*. They can replicate data objects in different variants (e.g., full-replication/selective replication), but have in general no further concepts to ensure conflict-free and durable local transaction processing. Gollmick [38] states that commercial mobile database systems are not aware of the transactional semantic defined by the application logic of a mobile application. Consequently, the local transaction manager of the mDBMS cannot decide which transactions to be executed could be potentially conflicting. Hence, only the separation of data objects during the replication could avoid conflicts. Consequently, all products use an *image-based* synchronization and therefore are limited to the inconsistency level of $n - 1$ (modification of at least one replication).

Nevertheless, since connectivity-awareness seems to have become a competitive factor for app vendors, we have found recently built frameworks such as Firebase [20] and

Smartface [23], and platform-specific built-in facilities such as Sync Adapter in Android [17]. However, these tools do not support either real offline transaction management or any particular conflict prevention. Developers must ensure conflict-free behavior of their mobile applications: that is, implement conflict-avoiding mechanisms after a manual identification of potential conflicts. Web-based mobile applications using HTML5 may also use replicated data objects stored locally in in-memory databases of the browser; the local storage, however, is very limited (up to 5 MB). Also none of these frameworks are combined with analysis or simulation capabilities for reasoning about the design.

2.3 MODELING OF CONNECTIVITY-AWARE MOBILE APPLICATIONS

A traditional development approach for mobile application development requires that all components of the generic extended client–server model mentioned before must be implemented manually. In particular, the identification of potentially conflicting processes and the separation of concurrently accessed data objects are not automated. Since we are heading toward an automated tool environment for the creation of online- and offline-capable mobile applications, many of the required tasks can be automated by using a model-driven development infrastructure. Hence, we introduce a domain-specific modeling language, which is the first essential part of the infrastructure for the model-driven development of online- and offline-capable mobile applications. However, the designed model-driven development infrastructure can also be used for the development of traditional online-only (client–server) or offline-only (standalone) applications.

2.3.1 Design Decisions

We use a component-based approach for our domain-specific modeling language (see Figure 2.10): an application model (app model) comprising a *data model* defining the underlying class structure, a *GUI model* containing the definition of pages and style settings for the graphical user interface, and a *process model* defining the behavior of a mobile application in the form of processes and tasks. The data model and the GUI model do not have a direct link. However, the process model includes dependencies to both submodels by referring to their elements.

2.3.2 Data Model

Figure 2.11 shows the core elements of the data metamodel. Mobile application modelers can create classes with typed attributes and connect them with different types of references

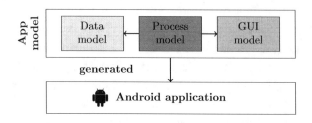

FIGURE 2.10 Structure of the domain-specific modeling language.

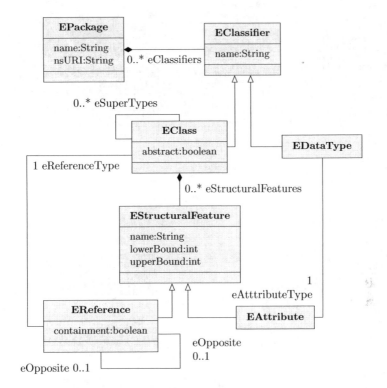

FIGURE 2.11   Metamodel (excerpt) used for data structures.

(e.g., inheritance, aggregation, association). Data models are not only used to generate the underlying object access facilities, but also to influence the presentation of data through the graphical user interface. Sub-objects, for example, result in a tabbed presentation of objects, attribute names are shown as labels (if not redefined), and attribute types define the appropriate edit elements such as text fields, checkboxes, and spinners. Class and attribute names are not always well-suited to be viewed in the generated mobile application—for example, an attribute name has to be a string without blanks and other separators, while labels in mobile application views may contain several words like "Account number". In such a case, an attribute may be annotated with the intended label.

2.3.3 Behavior Model

Figure 2.12 shows the metamodel for process models of mobile applications. This metamodel is influenced by the language design of BPMN [7] and (WS)-BPEL [5]. Since BPMN itself does not provide a built-in model for describing data structures, we have to reuse the data model described before. Thus, it is natural to adopt the required BPMN/(WS)-BPEL model elements to our own domain-specific modeling language. Many of the BPMN/(WS)-BPEL language elements have been removed (e.g., the error handling of (WS)-BPEL and the events of BPMN). The standard set of behavior constructs is extended by CRUD (create, read, update, delete) functionality in the data model, input/output facilities referring to the GUI model, platform-independent permissions, and CRUD privileges for which we do not find any adequate constructs in BPMN and (WS)-BPEL.

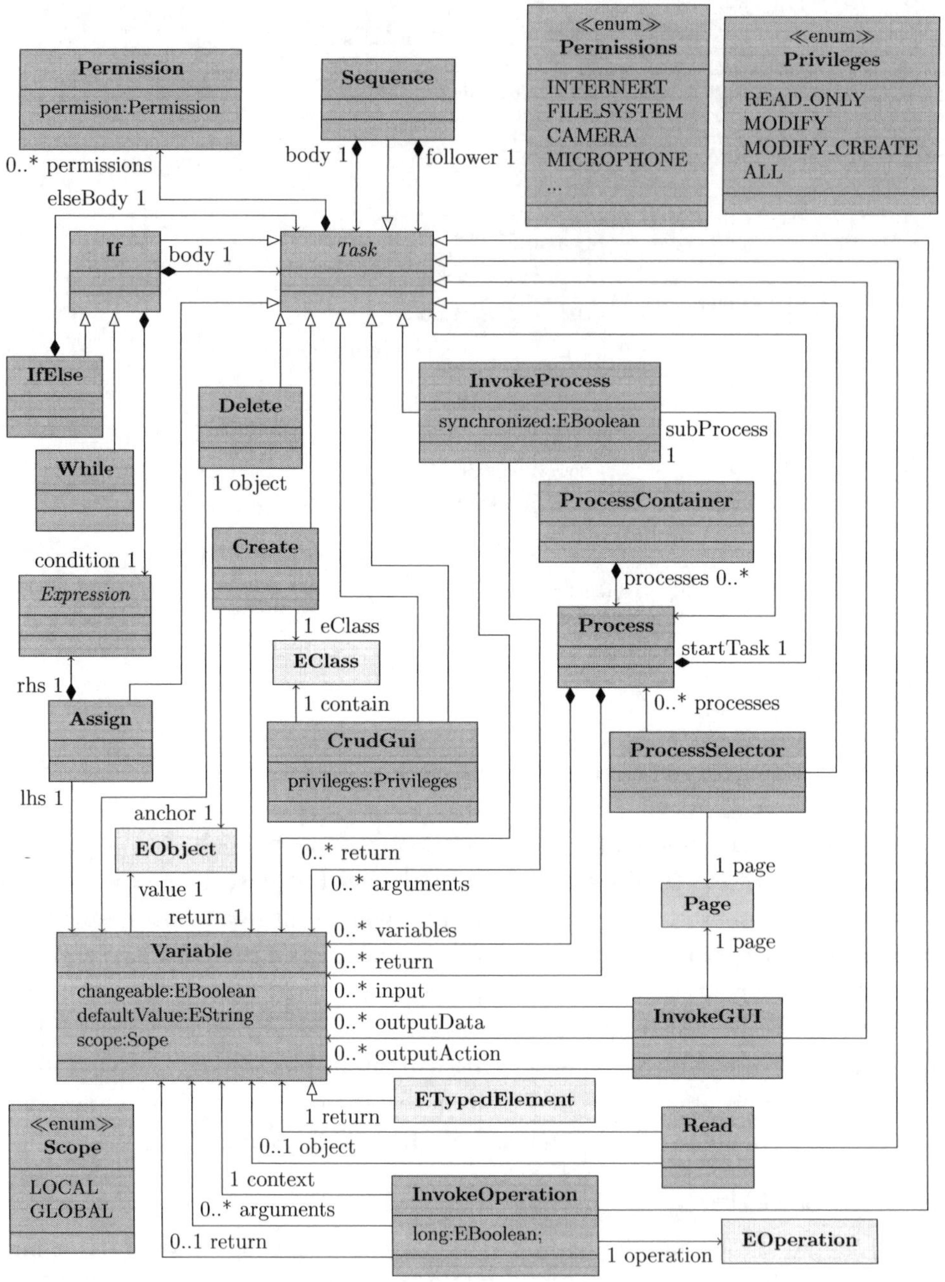

FIGURE 2.12 Metamodel for defining the behavior of mobile applications.

The main constituents of a process model are processes that may be defined in a compositional way. In particular, the modularity and reusability of existing processes requires minimal effort for process modeling. The model element *InvokeProcess* calls a sub-process. When invoking a process, the kind of invocation—*synchronous* or *asynchronous*—has to be specified. Long-running processes (e.g., processor-intensive or network-intensive processes) should be marked as *asynchronous*. These processes will be run in the background.

Each process has a name and several *variables* that may also serve as (return) parameters. A parameter is modeled as a variable with a global scope, contrary to locally scoped variables. The body of a process defines the actual behavior comprising a set of *tasks* ordered by typical control structures; it is potentially equipped with permissions.

Permissions indicate the required rights (e.g., network, file access, Global Positioning System [GPS]) of the mobile applications. There are a several predefined tasks covering basic CRUD functionality on data objects (e.g., *Create*, *Read*, *Update*, and *Delete*), control structures (e.g., *If*, *If-Else*, and *While*), the invocation of an external operation (*InvokeOperation*) or an already defined process (*InvokeProcess*), and the view of a page (*InvokeGUI*). While the task *CrudGui* covers the whole CRUD functionality with corresponding views, *Create*, *Read*, and *Delete* just cover single internal CRUD functionalities.

Privileges can limit object access (e.g., Read-only, Modify, and Modify & Create) of the element *CrudGui*. An *InvokeGUI* task refers to a page defined in the GUI model. The *ProcessSelector* points to all processes that are to be available in the main menu of the mobile application.

2.3.4 Graphical User Interface Model

The metamodel for GUI models is shown in Figure 2.13. Different kinds of graphical user interfaces are modeled by different kinds of pages (e.g., *ViewPage*, *EditPage*, and *MapPage*). Each of these pages has a predefined (generic) structure of graphical user interface components and follows a specific purpose. For example, the purpose of *EditPage* is to edit attributes of an object (e.g., Address). Our GUI model reuses the existing data model that holds a description of the data objects. In addition, mobile application modelers can set style and presentation properties. The different style setting elements of our GUI model provide this aspect of presentation: these elements influence the style of *Pages* (*PageStyleSetting*) in general and in particular of *Menus* (*MenuStyleSettings*), *Lists* (*ListStyleSettings*), and *Selections* (*SelectionStyleSettings*). A dialog sub-model (cf. Trætteberg [69]) does not exist in our modeling approach. This conversational aspect of the graphical user interface is covered by pages that implicitly contain the necessary dialogs.

Example (Modeling the Payment App)

The simple payment app is modeled as follows—the *data model* in Figure 2.14a mainly contains a class *Account* that models a banking account. It comprises an attribute representing the current balance. It can be checked by using *getBalance()* and modified by the operations *deposit()* and *withdraw()*. The two modifying operations need an amount that should be added to or subtracted from the balance. The operation *withdraw()* requires the availability of sufficient funds to cover the required amount. Its return variable is set to *true* if covered and to *false* if not. The main process of the *process model* in Figure 2.14b is a process selector which refers to all the available processes. The mobile user may invoke the processes *Withdraw* (Figure 2.14c), *Deposit* (Figure 2.14d), or *GetBalance* (Figure 2.14e). The process *GetBalance* invokes the operation *getBalance()* and displays the value. The process *Deposit* requests the

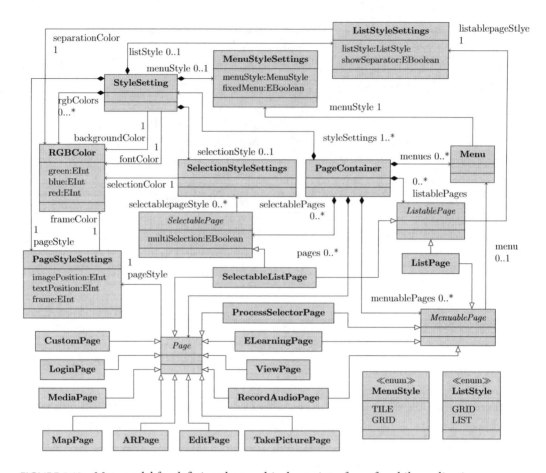

FIGURE 2.13   Metamodel for defining the graphical user interface of mobile applications.

amount of money that should be added and then invokes *deposit()*. The process *Withdraw* also requires the amount of money to be withdrawn, invokes *withdraw()*, and displays whether the transaction can be carried out or not. A simple GUI can also be modeled and generated (see Figure 2.14f). The mobile application shown in Figure 2.15 is generated by our framework for the model-driven development of mobile applications presented in [73]. The architecture of this mobile application follows a traditional client–server architecture; that is, the mobile application is not operable without a network connection.

2.3.5 Modeling of Connectivity-Awareness

The preceding section shows the general modeling activities for a mobile application. The modeling elements are already useful for the creation of online- and offline-capable mobile applications because a model analysis can find potential conflicts that must be handled in offline situations through the description of the data and behavior. Thus, connectivity-awareness modeling does not necessarily have to be performed in an explicit way. However, some information about the online- and offline-capable behavior must be given explicitly. First, the data access of custom operations (part of the data model) must be declared.

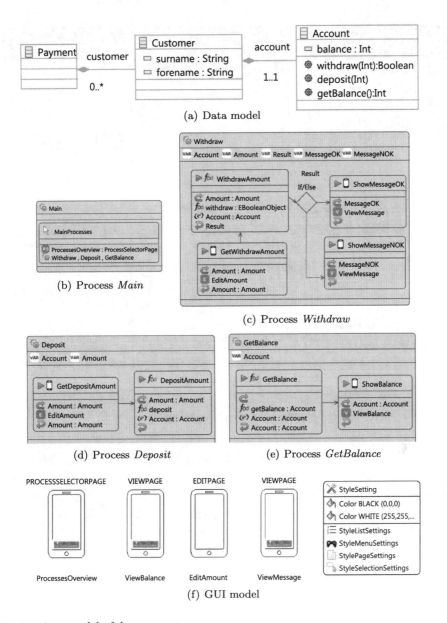

FIGURE 2.14   App model of the payment app.

Second, as shown in Section 2.2, mobile transaction models might restrict the data access due to the split of custom objects or summable objects. Considering this, mobile application developers must declare by explicit annotations which classes of the data model should be handled in a connectivity-aware way.

2.3.5.1 Extract Data Object Access

In order to apply our read/write-based conflict definition, conflict-relevant information (i.e., the read/write access) must be extracted from the app model. Extraction rules deliver data object access of the different types of tasks. A task can have an Action:Read

(a) *Main*   (b) *GetBalance*   (c) *Deposit*

(d) *Withdraw*

FIGURE 2.15   The payment app (online-only version).

or `Action:Write` on an object or its attribute(s). Moreover, to distinguish between different conflict levels (e.g., consistent read and write, inconsistent read/consistent write), read operations within a conditional expression get a `Condition:Read` instead of an `Action:Read`. The conflict-relevant information of any single task is joined to a *conflict set* that belongs to a process. Table 2.1 shows examples of the main modeling elements of the domain-specific modeling language and the conflict information that is extracted. We will emphasize that the extraction of data object access is fully automated—that is, no extra effort for explicating data access is needed, as long as mobile application modelers use standard modeling elements.

2.3.5.2 Explicating Data Object Access

To apply conflict analysis—determining whether a custom operation is potentially involved in a conflict or not—all custom operations must declare their data accesses. We use the same language features as in the prior section to denote the data object access of custom operations:

    i. An operation has an `Action:Read` (`Action:Write`) relation with an attribute or object if it returns (writes) this attribute or object.

    ii. An operation has a `Condition:Read` relation with an attribute or object if it reads this attribute or object as part of a condition.

TABLE 2.1   Extraction of Conflict Relevant Information

Model Element (Task)	Conflict Information
 InvokeGUI (ViewPage)	**Action:Read(Account):** The task ShowBalance (InvokeGUI) shows an account on the page ViewBalance (ViewPage). The account cannot be modified.
 InvokeGUI (EditPage)	**Action:Read/Write(Account):** The task ChangeBalance (InvokeGUI) shows an account on the page ChangeBalance (EditPage). The account can be modified.
 Create	**Action:Write(Customer.account):** The creation of accounts affects the class Customer because new accounts must be part of a customer.
 Read	**Action:Read(Account):** The task ReadAccounts (Read) returns all data objects of the type Account.
 Delete	**Action:Write(Customer.account):** The task DeleteAccount (Delete) deletes an account. Even the account is not accessed the customer object is affected.
 CRUDGui (All)	**Action:Read(Account), Action:Write(Account, Customer. account):** The task CRUDAccount (CrudGui) provides reading and modifying accounts. Moreover, the Creation and deletion of accounts affects the class Customer.
 Assign	**Action:Write(Account.balance):** The task Withdraw (Assign) assigns a new value to the attribute balance.
Account.balance-Amount>=0 If/Else Conditional tasks (If, If/Else, and While)	**Condition:Read(Account.balance):** The conditional task (If/Else) reads the attribute balance (and the unbound type Amount) and branches to the appropriate path.
...	...

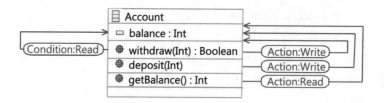

FIGURE 2.16   Access modes of the class *Account*.

Example (Explicating Data Object Access of Custom Operations)

Listing 1 shows the operation *withdraw()* of the class *Account*. This operation reads the attribute *balance* to check whether the account is covered. Hence, it has a `Condition:Read` relation with *balance*. It writes the attribute *balance* inside the body of the conditional statement, leading to an `Action:Write` relation with *balance*. Figure 2.16 shows the data object access declarations for all operations of *Account* by using newly introduced language features for explicit declarations of data object access.

Listing 1 Custom operation *withdraw()*

```
public boolean withdraw(int amount){
  if ( balance  - amount >= 0){
       Condition:Read
    balance = balance - amount;
    Action:Write
    return true;
  } else {
    return false;
  }
}
```

We will see later how these declarations are used during the model analysis step.

2.3.5.3 Restricting Data Object Access

As explained already in Section 2.3.4, the offline transaction processing state of the working model needs information about which classes of the data model should be handled in a connectivity-aware manner, that is in a pre-processed way. Thus, the <ca> annotation is already a model element for declaring the connectivity-aware behavior. Depending on the context (i.e., positioned at the attribute or class label—cf. Figures 2.5 and 2.7) of the <ca> annotation, it denotes that an attribute is an aggregated value as part of an summable object or a class represents a set of custom objects. Accordingly, the Escrow or the Keypool method will be used. Furthermore, the #AllocObjects value denotes how many fragments of an aggregate or custom objects will be allocated to mobile clients.

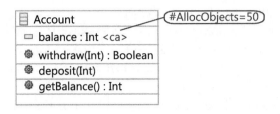

FIGURE 2.17 Restrictions of the class *Account*.

Example (Application of Mobile Transaction Models)

Figure 2.17 shows a snippet of an annotated data model for our simple payment app. The annotation <ca> (connectivity-aware) of the attribute *balance* indicates that this object may be split and allocated to mobile clients. Thus, mobile clients may share *Account* objects, particularly its attribute *balance*. The parameter *#AllocObjects* denotes how much of the aggregate *balance* should be allocated (i.e., distributed among mobile clients). For example, assuming the attribute balance has the value of 100, *#AllocObjects* is 50, and the aggregate is distributed among two clients. Then each mobile client receives an amount of 25 (cf. Figure 2.6).

2.4 MODEL ANALYSIS, EVALUATION, AND PROTOTYPE GENERATION

Having presented the domain-specific modeling language and the accompanying model editor (i.e., its visual syntax) for modeling app models with connectivity-aware annotations, we can now go on to consider model analysis, design evaluation, and, finally, prototype generation.

2.4.1 Analyzing Data Object Access and Conflicts

The purpose of model analysis is to evaluate how many processes are potentially in conflict and which conflict-free configurations exist. For example, an online- and offline-capable mobile application requires replicated data objects in the mobile device on which it is running. The replicated data objects may be modified locally while the mobile device remains disconnected. In parallel, another mobile client can potentially do the same with its replicated data objects. This indicates a concurrent access to shared data objects, but traditional online conflict detection cannot be applied due to the network partition of mobile clients [27].

Hence, the following question emerges: which conflicts may occur when clients try to synchronize their modified copies and how can these conflicts be prevented or resolved? Thus, we first introduce an appropriate definition for the term conflict. Since this definition is based on read and write access to data objects, the annotations described in Sections 2.3.5.1 through 2.3.5.3 are utilized in the analysis step. Finally, conflict analysis can run automatically; it delivers a conflict matrix as well as conflict-free subsets of processes.

2.4.1.1 Conflict Definition and Conflict Levels

The traditional conflict definition for online transaction systems is based on the *page model* [27] and considers conflicts between read and write operations of different transactions. The page model and its read-/write-based conflict definition are very restrictive and may prevent execution of transactions that are actually conflict-free. For example, the process *Deposit* reads and writes the balance of an account and thus is in conflict with itself, assuming that the single steps of the transactions are interleaved. However, regarding the process *Deposit* as an atomic increment operation of the account balance, parallely executed deposit transactions have the same effect as any serial execution because the increment is a commutative operation. Hence, the *object model* was introduced (cf. Weikum and Vossen [77]) as a generalization of the *page model*. It defines conflicts based on noncommutative object operations instead of low-level read/write operations. This fits very well with the requirements given in the mobile domain. Transactions that are performed offline on replicated data objects must be *repeatable* on the primary copy (located at the server) in any order because the mobile clients will reconnect and synchronize in an arbitrary order. An offline transaction is repeatable only if it always returns the same value, that is, if it fulfills the *return value commutativity*, and has the same effect, that is, if it satisfies the *state commutativity* when reexecuted on the primary copy.

Example (Return Value Commutativity Conflict)

We assume that two clients access a banking account with an initial balance of $50. One of them requests the account balance while being offline. The process *GetBalance* delivers the last consistent value of *balance* being $50. In the meantime, the other client deposits $50 in an offline context (*Deposit*). Hence, the return value of the process *GetBalance* is not the same on the primary copy (assuming an arbitrary order of synchronization/reintegration), because it may deliver $100 instead of the locally returned value of $50. The processes *GetBalance* and *Deposit* are conflicting.

Example (State Commutativity Conflict)

In this scenario, we assume that a bank account is accessed offline by a mobile client and online from an ATM. Its initial balance is $50. The mobile client withdraws $30 through the offline transaction *Withdraw* (e.g., transferring money via NFC) and this transaction is performed on a local copy. It can be carried out locally because the amount paid is covered by the local balance of $50. The local copy of the balance is $20 now. In the meantime, the mobile user withdraws $30 from an ATM. This is an online transaction accessing the primary copy. It is checked whether the account is covered using the primary copy (which is still $50) and debits the requested amount. Finally, the primary copy has a value of $20. While synchronizing the offline transaction, a conflict occurs because the amount has to be covered. In particular, the condition (blance − amount ≥ 0) is violated. The process *Withdraw* is in conflict with itself, because local executed *Withdraw* transactions will not preserve state commutativity.

Return value commutativity conflicts (i.e., reading inconsistent values) are tolerable in some scenarios, whereas state commutativity conflicts (i.e., lost updates) should usually be avoided. Similar to the ANSI/ISO SQL [24,26] transaction isolation levels, we propose *weak* and *strict* versions of the return value commutativity and state commutativity respectively. Thus, developers can use a fine-grained definition for the required conflict-freeness.

Two offline transactions (not necessarily of the same process type) are called *weakly return value commutative* if their return values may differ when being reprocessed on the primary copy. This happens if other clients have changed the primary copy in the interim period. Otherwise, they are called *strictly return value commutative*.

Two offline transactions (not necessarily of the same process type) are *weakly state commutative* if the transformation of one state to another one may fail. This happens if other clients have changed the primary copy in the interim period and hence the execution condition of the failed transaction is no longer satisfied. They are *strictly state commutative* otherwise.

The toleration of conflicts depends strongly on the application scenario of the mobile application. We define four relevant conflict levels in order to define conflict strategies. The conflict strategy is a global property: that is, all transactions have to follow the same conflict level:

Level C1 (Conflict-allowing) requires weak return value commutativity and weak state commutativity.

Level C2 (Conflict-avoiding) requires weak return value commutativity and strict state commutativity.

Level C2-MTM (Conflict-avoiding using mobile transaction models) requires weak return value commutativity and strict state commutativity, except state-changing transaction which are performed on summable or custom objects. More precisely, transactions are C2-MTM conflicting if they are C2-conflicting and will not use a mobile transaction model.

Level C3 (Conflict-prohibiting) requires strict return value commutativity.

Note that weakness or strictness of state commutativity (Level C4) does not matter while requiring strict return value commutativity, because state-changing operations, that is, write operations, are not allowed. However, if the transaction set contains only writing transactions that did not previously read any data object, the transactions are always strictly state commutative because the state-changing transaction does not requires a particular state. The conflict level C3 is equal to a standard conflict definition, for example, used in online-only client–server architecture.

Conflicting processes are identified as follows:

- The conflict level C1 does not contain conflicting processes because any kind of conflict is accepted by this conflict level.

- Two processes are C2-conflicting if one has a `Condition:Read` and the other has an `Action:Write` relation with the same attribute/class in its *conflict set*.

- Two processes are C2-MTM-conflicting if one has a `Condition:Read` and the other has an `Action:Write` relation with the same attribute/class in its *conflict set*. Additionally, the attribute/class has no <ca> annotation.

- They are C3-conflicting if one has an `Action:Read` or a `Condition:Read` and the other one has an `Action:Write` relation with the same attribute/class in its *conflict set*.

Transactions that consist only of `Action:Write` relations (but no `Action:Read` or `Condition:Read`) are named *blind writing* transactions [27]. These transactions are return value commutative (because they have no return value) and strict state commutative (because their execution depends on no particular state).

Example (Conflict Levels)

Considering our payment app, all processes fulfill C1. Level C2 dismisses the process *Withdraw* because the state condition is not fulfilled for this process. Hence, *Withdraw* is self-conflicting. Level C3 dismisses the processes *Withdraw* and *Deposit*, or *GetBalance*, because either the processes *Withdraw* and *Deposit* violate the required strict return value commutativity, or *Withdraw* and *GetBalance* violate the state-independent write-only transaction set of the application.

If a mobile application has either read-only processes or processes with a write access, albeit without a conditional read access, it may gain most of the connectivity-awareness as its processes are conflict-free. All other mobile applications have to accept or resolve conflicts. We can identify different kinds of mobile applications for which conflict levels are beneficial. The conflict-allowing level, C1, is recommended for mobile applications that realize an *information system* where the data flow would mostly be from the server to mobile clients. Conflict levels C2 and C3 are rather suitable in *transaction systems* where the data flow is bidirectional between the server and mobile clients. However, the appropriateness of a conflict level is mostly determined by the number of users and data objects, that is by the conflicts that actually occur.

2.4.1.2 Running the Model-Driven Static Conflict Analysis
Input of the analysis step: According to Figure 2.1, the input of the analysis step is an app model equipped with connectivity-aware annotations. As shown in Figures 2.16 and 2.17, only the data model might be explicitly equipped with annotations, but the whole app model (with process model) is considered.

The tasks of all processes are analyzed for conflicts based on the given conflict definition (cf. Section 2.4.1.1). All conflicting processes are identified based on their *conflict sets*. The model analysis can automatically recognize potential conflicts for all completely modeled processes (cf. Section 2.3.5.1). For customized operations that are invoked inside modeled processes, modelers must manually explicate the data object access of these operations by using additional language features as shown in Section 2.3.5.2. Moreover, the model analysis can also consider the restrictions of data objects as described in Section 2.3.5.3.

Three out of the four processes of the payment app use customized operations. We have chosen this example to explore all kinds of conflict levels in a single mobile application.

Example (Conflict Analysis)

For our payment app, the result of the conflict analysis is shown in Table 2.2. It is a conflict matrix, where "✓" marks show conflict-free processes and "×" marks show the conflicting ones for the corresponding levels. We assume for this example that the modeler does not restrict data object access, that is does not apply mobile transactions models.

A conflict-free configuration can be constructed by removing (i) self-conflicting processes and then (ii) processes involved in a conflict until the set is conflict-free. Table 2.3 shows the removed self-conflicting process, namely *Withdraw* (Case i). The remaining processes—*Deposit* and *GetBalance*—are still C3-conflicting. Hence, the inhibition either of the process *Deposit* or of the process *GetBalance* delivers a conflict-free application variant.

Output of the analysis step: As shown in Table 2.2, the analysis step delivers a conflict matrix that at first shows all potentially conflicting processes. From this conflict matrix, a conflict-free configuration for a particular conflict level can be derived (cf. Table 2.3) which will be part of the input for the simulation step.

Example (Connectivity-Aware Payment App)

Figure 2.18 shows the generated online- and offline-capable payment app that applies the results of the analysis step at runtime. The mobile application provides all available processes at runtime in accordance with the connection context and the selected

TABLE 2.2   Conflict Matrix (C1, C2, C3)

Processes	Withdraw	Deposit	GetBalance
Withdraw	$(\checkmark, \times, \times)$	$(\checkmark, \times, \times)$	$(\checkmark, \checkmark, \times)$
Deposit	$(\checkmark, \times, \times)$	$(\checkmark, \checkmark, \checkmark)$	$(\checkmark, \checkmark, \times)$
GetBalance	$(\checkmark, \checkmark, \times)$	$(\checkmark, \checkmark, \times)$	$(\checkmark, \checkmark, \checkmark)$

TABLE 2.3   C3 Configuration Variants

Processes	~~Withdraw~~	Deposit	GetBalance
~~Withdraw~~	~~×~~	~~×~~	~~×~~
Deposit	~~×~~	✓	×
GetBalance	~~×~~	×	✓

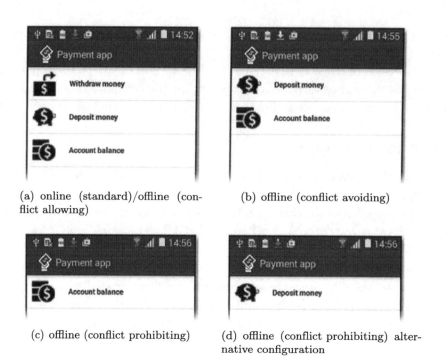

(a) online (standard)/offline (conflict allowing)

(b) offline (conflict avoiding)

(c) offline (conflict prohibiting)

(d) offline (conflict prohibiting) alternative configuration

FIGURE 2.18 The generated payment app (online- and offline-capable).

conflict level. Choosing the conflict-allowing level C1, all processes can be offered (as in the online mode, Figure 2.18a). Conflict level C2, however, requires the removal of at least one conflicting process, which, in this case, happens to be *Withdraw* (see Figure 2.18b). To reach conflict level C3, two processes have to be removed to prohibit conflicts—these are *Withdraw* and *Deposit* (see Figure 2.18c) or *Withdraw* and *GetBalance* (see Figure 2.18d) in this case. We recommend the conflict-avoiding level C2 for the payment app: it ensures that the balance is always funded because money cannot be withdrawn without checking that sufficient funds are available. In turn, mobile users can check the balance and deposit money even when offline.

However, the conflict level is not only driven by the requirements at a mobile application (with respect to conflict-freeness), but it also depends on the number of actual conflicts taking place in the system. Developers are interested in understanding how throughput would improve if such conflicts were accepted. Hence, a simulator can support developers by estimating the throughput for different conflict levels.

2.4.2 Model-Based Simulation and Evaluation

As Tables 2.2 and 2.3 show, conflict analysis delivers conflict-free subsets of processes by discarding conflicting processes (cf. Figure 2.18). The inhibition of conflicting processes is effective, but it may pose the following question: How many conflicts would there be if potentially conflicting processes are not inhibited? For example, an online-only architecture might be generally nonoperable in an offline context, but an online- and offline-capable mobile application might be too restrictive and hence nonoperable due to process-starvation.

The actual throughput of mobile applications depends on several factors such as the number of mobile users and data objects. Other factors are the activity, connectivity, and behavior of mobile users. The structure of data objects (e.g., loosely coupled or strongly related) also affects the throughput. We have observed that these runtime factors can considerably influence the number of actual conflicts in both positive and negative ways. For example, an app model with many potentially conflicting processes might be uncritical when only a few mobile users of low activity are involved. In contrast, an app model with less conflict potential might be critical when a lot of highly active users act on a small set of data objects.

2.4.2.1 Dynamic Conflict Analysis by Simulation

To support developers better in evaluating the results of static conflict analysis, the proposed design process contains a simulation step. Based on the app model and different sets of conflict-allowing, conflict-avoiding and conflict-prohibiting processes, developers can predict and compare throughput of their mobile app on the basis of a dynamic conflict analysis.

Input of the simulation step: First the simulator requires the app model to determine the throughput for a standard online-only architecture. All defined processes are considered during this simulation run. Further on, the simulation system also requires a set of conflict-free processes (as calculated by the prior analysis step; cf. Table 2.3) to determine the throughput for a conflict-free, online- and offline-capable architecture. During this simulation run, the simulation system considers only the limited, conflict-free subset of processes. Moreover, values for all independent simulation variables have to be set. They are shown in Table 2.4. *#Users* denotes the number of simulated mobile clients. *Activity* denotes the average activity level of all mobile users. A level of 50 means that, on average, half the mobile users are active in each iteration. *Connectivity* denotes the average connectivity of all users in each iteration. By setting the number of objects (*#Objects*) of each class, the simulation system constructs a corresponding instance of the data model used in the simulation. Developers can also load an existing instance of the data model to simulate on the basis of the data taken from a running system. This is indicated by the custom instance (row *#Objects*) in Table 2.4.

TABLE 2.4   Independent Simulation Variables

Name	Description	Domain
App model	Definition of data structure, behavior and GUI.	DSML
Subset of processes	Conflict-free C1-, C2-, C2-MTM-, and C3-subsets.	$\mathcal{P}$ (modeled processes)
#Users	Number of mobile clients.	1 ... 10,000
Activity	Average activity level of users.	0 ... 100
Connectivity	Average connectivity level of users.	0 ... 100
#Objects	Number of objects or aggregate size for each class.	1 ... 5,000/ Cust. instance

Abbreviation: DSML, domain-specific modeling language.

TABLE 2.5   Dependent Simulation Variables

Name	Description
#Processed_Client_n	Number of transactions processed by Client n.
Throughput	$\sum\limits_{n=1}^{\#Users} \#Processed\_Client\_n$

Output of the simulation step: In contrast, the simulation system delivers the so-called dependent simulation variables as output which is shown in Table 2.5. The number of transactions processed by Client n is given by *#Processed_Client_n*. The overall system throughput is the sum of these values. These dependent simulation variables are determined for all conflict levels of the online- and offline-capable architecture as well as for the online-only architecture.

2.4.2.2 Running the Model-Based Simulation

Based on the app model and input parameters, that is, the independent variables, the simulation system determines the throughput of the desired mobile application. With the exception of the app model, not all input parameters must be set. For example, the investigation of how the throughput varies due to the different connectivity levels of mobile clients may leave this parameter (connectivity) uninitialized. The simulation system selects values for the uninitialized parameters in accordance with their corresponding domains (cf. Table 2.4). Every run passes by default 10,000 simulation iterations to get stable results.

Example (Predicting the Throughput of the Connectivity-Aware Payment App)

We apply the simulation system to the model of our simple payment app, which has been shown in Figures 2.14, 2.16, and 2.17. Figure 2.19 shows the results of the simulation with 500 mobile clients and 100 accounts. The results are normalized according to the maximum throughput. The *Online* graph represents the throughput

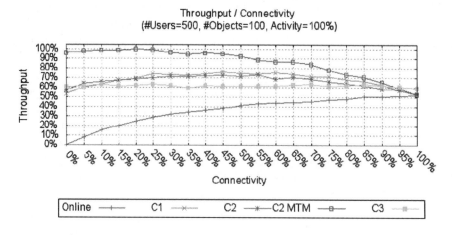

FIGURE 2.19   Throughput of the payment app (Configuration 1).

for the online-only architecture while using a standard transaction model and conflict definition (cf. Weikum and Vossen [77]). At the best connection level (100%), the overall system reaches a throughput of 50%. It is not higher than that owing to the occurrence of conflicts. It deteriorates as the average connection level of the clients (users) declines. At the worst level (0% connectivity), there is no throughput (0%). The *Offline (C1/C2/C2 MTM/C3)* graphs show the throughput at different conflict levels. While even the C3 level, which is by far the most restrictive level, provides a higher throughput than the online-only variant, the best variant is actually shown by graph C2 MTM, which denotes the C2 level using mobile transaction models in addition. The throughputs of C1, C2, and C3 variants are almost equal for the simulated configuration. Another tested configuration, shown in Figure 2.20, demonstrates that a conflict-prohibiting conflict level (C3) deteriorates the throughput when conflicts occur rarely. In this case, it is advisable to select the conflict-allowing level (C1). Besides, the use of a mobile transaction model will provide only marginal improvements.

2.4.2.3 Design of the Simulation System

The purpose of the simulation system is to predict how many conflicts would occur if the designed online- and offline-capable mobile application were implemented. Thus, the simulation system generates a set of mobile users, a functional implementation of the data model and the behavior model, and a conforming set of data objects. During a simulation run, mobile users execute the processes (transactions) generated by a customizable transaction generator. The simulation system logs every transaction, even if it fails, that is executed offline and online, and finally delivers a global history. This history is analyzed, and eventually, the throughput is delivered, as shown in Figures 2.19 and 2.20.

To conclude, the simulation system generally shows whether an online- and offline-capable architecture is preferable. Moreover, it shows which throughput could be achieved by accepting different conflict levels. Finally, it predicts the throughput for using mobile transaction models.

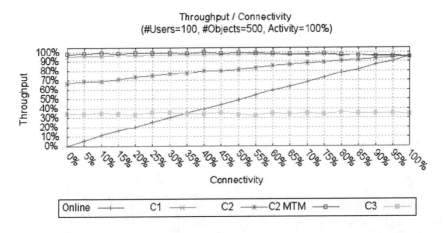

FIGURE 2.20 Throughput of the payment app (Configuration 2).

2.4.3 Prototype Generation

While the first steps of the design process focus on the analysis of an application's design based on its online and offline capability, the final step is applied to generate a software prototype of the designed mobile application. However, the results of the analysis and the simulation steps are decisive factors in deriving an adequate architecture for the desired mobile application. Based on the configuration of the code generator, a mobile application with an online-only architecture (cf. Figure 2.15) or an online- and offline-capable mobile application (cf. Figure 2.18) can be generated.

Input of the generation step: The generation step basically requires an app model. To generate a connectivity-aware mobile app, the app model must be equipped with the above-mentioned annotations that either declare the data access (cf. Section 2.3.5.2) or restrict it (cf. Section 2.3.5.3). The simulation results are not directly processed by the generator because it seems more appropriate to let the developers decide the conflict level to be applied in the resultant prototype. The developers can select between Online, C1, C2, C2 MTM, and C3.

Output of the generation step: The output of the generation step is a native mobile Android application that has either a simple client–server architecture or a connectivity-aware architecture. Based on the annotations made, for example, <ca>, the corresponding mobile transaction models will be applied automatically in the generated prototype.

Based on the our existing framework [73] for the model-driven development of mobile applications with a client–server architecture, we introduce the generation of connectivity-aware applications following the generic extended client–server architecture (see client *GEC* in Figure 2.8). Hence, the existing code generator was extended to generate mobile applications containing a local transaction manager (TM) and a local database. The following functions were realized based on the working model of the TM (cf. Figure 2.9):

Initial setup: Given the data model, the model-object mapping framework Teneo [15] and the object-relational mapping framework Hibernate [48] allow us to set up a relational database scheme and to persist model instances in a server-located database.

Online transaction processing: Since Hibernate is a certified JPA provider [33,43], it includes transaction session management that can be reused to handle the online transaction processing.

Replication: Data replication is realized by loading the model instances from the database (server), detaching them from the online session, and storing them locally on the mobile devices.

Offline transaction processing: Offline transaction processing follows the selected conflict level. All transactions are logged for later synchronization/reintegration and will be executed on the server when the mobile devices are online again.

Synchronization and reintegration: Since we use a process model that defines the behavior in an explicit way or define custom operations by attached program code in the data model, the mobile applications can realize a transaction-based reintegration (cf. [66]).

By logging the object identifiers of accessed data objects and performed operations (including parameters) in the offline context, a transaction can be repeated on the server in the online mode.

To sum up, model-driven generation step allows the prototypical generation of a connectivity-aware application. By understanding the concepts presented here, mobile application developers can generate online- and offline-capable applications that might serve as prototypes for further developments.

2.4.4 Case Example: Air-Quality Application

In this section, we aim to answer the following question: Is our design process for connectivity-aware mobile applications useful and applicable? To address this question, we investigate whether our design process can be easily applied by mobile application developers to create software prototypes with a connectivity-aware architecture.

We present a case example that demonstrates the model-driven reengineering of an existing air-quality application with the goal of making it online- and offline-capable. The reengineering of this app results in a prototype that should demonstrate the applicability and suitability of our model-driven design process. The air-quality application—called WeSense—was developed in its initial form in the context of the EU project GameBus, which was coordinated by the Technical University of Eindhoven (TU/e), Netherlands. It provides real-time air pollution data for the cities of Eindhoven, Breda, and Helmond. The measurements (together with their corresponding locations) are plotted on a Google Map (cf. Figure 2.21a). In addition, the application provides the opportunity to comment on air-quality values and take pictures of critical locations (cf. Figure 2.21b). User comments are displayed along with the air pollution data on the map. The drawback of WeSense is that it needs a permanent network connection to update its data (cf. Figure 2.21c).

For the first step toward the creation of a prototype, an app model comprising a data-, process-, and GUI-model is created. Regarding the design process shown in Figure 2.1, the analysis of the WeSense application does not return any conflicting processes because the user comments can be added without any conflict and the air-quality data is always static. The simulation shows that such an online- and offline-capable application of read-only character enables a success rate of 100% in offline situations, while the success rate of the former online-only WeSense application is connectivity-dependent. Hence, a mobile application with connectivity-aware architecture is generated (cf. Figure 2.21d through f) and provides online and offline access. However, simulations for other kinds of mobile applications (e.g., an AppShop) show that the average improvement of an online- and offline-capable architecture compared to an online-only architecture can be marginal depending on circumstances. This is, in particular, the case if the mobile application contains conflicting processes, requires a high conflict level (C3), and the ratio between mobile users and data objects is low. In such cases, the simulation step reveals that the mobile application gains the most throughput using a traditional client–server architecture.

Having demonstrated the applicability of our model-driven design process for a real-life mobile application, we will compare the development time using different development approaches. The

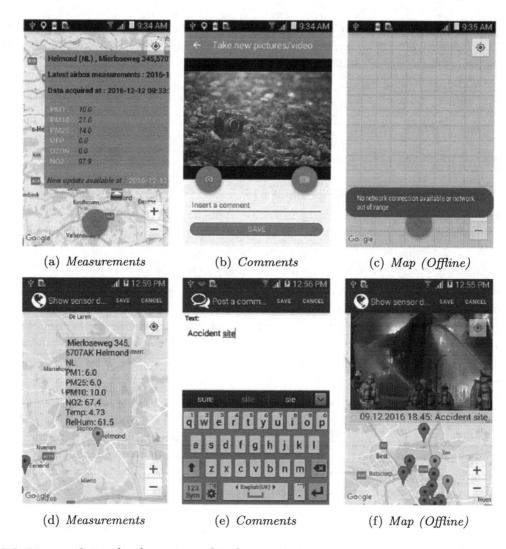

(a) *Measurements* (b) *Comments* (c) *Map (Offline)*

(d) *Measurements* (e) *Comments* (f) *Map (Offline)*

FIGURE 2.21 Original and reengineered application *WeSense*.

model-driven reengineering of the prototype application took 128 hours, which includes the review of the existing application and the back-end, including its application programming interface, the modeling of the app model, and the indoor and outdoor tests. In contrast, the former WeSense application development project produced nearly the same workload, but the resulting application is not connectivity-aware. Hence, we demonstrate that the presented model-based design process is both applicable and beneficial regarding the development effort.

2.5 RELATED WORK AND CONCLUSION

There are several approaches that support a model-driven design of context-aware mobile applications, some of which directly support the design of connectivity-aware mobile applications. Hence, we discuss how modeling of context-awareness is possible, which contexts (e.g., platform, user, device, location, and connectivity contexts) are supported in general, and which concerns of the mobile applications can be adapted (e.g., data, behavior, GUI) for these contexts. Model-driven design approaches that support the development of context-aware mobile applications are proposed by Serral et al. [62], Ceri et al. [30], Escolar et al. [36], and Kapitsaki et al. [42]. The approach of Serral et al. and Kapitsaki et al. requires explicit modeling of the context-aware behavior, while the other examples, including ours ([67,70]), use an implicit or mixed approach to model context-awareness. While connectivity-awareness is only dealt with in the approach of Escolar et al. contexts such as user, device, and location are supported by most of approaches mentioned. Since connectivity-awareness has a major impact on the architecture of mobile applications it is only provided rarely and in rudimentary form.

In the aspect of online and offline capability, the approach of Escolar et al. has a lot of similarities with ours: for example, both realize architectural components such as replication and synchronization/reintegration. The solution of Escolar et al. presents a detailed configuration level as it considers how data objects should be managed; however, it does not provide model analysis and cannot deal with conflicting processes appearing in multiuser transaction systems. As a result, the approach of Escolar et al. provides only connectivity-specific adaption of data objects (i.e., keeping data objects in sync across different connection states), instead of connectivity-specific behavior, that is, limiting conflicting processes in offline contexts. Moreover, the web-based solution of Escolar et al. can be used if a mobile application neither requires complex hardware access to the sensors of the device nor supports conflicting multiuser access to data objects.

2.5.1 Conclusion

To summarize, in this chapter we identify and present software architectures that are useful for implementing online- and offline-capable mobile applications. We provide a domain-specific modeling language tailored to the domain of mobile applications. This allows modeling of connectivity-aware mobile applications. A model analysis step identifies conflicting processes, which will be managed in a certain way while a mobile device is offline. Model-based simulation predicts the number of transactions expected for an application design, that is, an app model. Furthermore, a code generator can be used to create runnable software prototypes of online- and offline-capable mobile applications. The software prototypes can be used for experiments and as a starting point for further development activities. Thus, the effort of realizing connectivity-aware mobile applications can be reduced. The examples demonstrate that the number of overall transactions can be increased because disconnected mobile clients are no longer inoperable.

We learned the following lessons during the development and use of our model-driven design approach for connectivity-aware mobile applications:

- Model-driven development provides a high abstraction level of development: application models describe data structure, behavior, and the graphical user interface of a mobile application in a concise and platform-independent way. It is promising to declaratively specify which data objects should be handled in a connectivity-aware manner (e.g., replicated and synchronized) rather than being coded in a low-level way.

- Model-driven design helps in developing adequate support for offline operation: mobile application developers do not have to identify the (conflicting) processes and necessary data objects for offline operation of mobile applications. The model-driven infrastructure provides this task and automatically generates mobile application with adequate behavior. Moreover, we show the behavioral correctness of generated mobile transaction systems using Colored Petri Net (CPN) models and corresponding model checkers [45].

- Model-driven development provides high-quality software prototypes: consequently, the model-driven infrastructure applies design patterns for managing transaction and data in a connectivity-aware mode. The generated software prototypes follow a widely accepted software architecture.

- Model-based simulation helps mobile applications developers to decide whether an application design is beneficial to operate in offline contexts.

Our model-driven design approach for connectivity-aware mobile applications supports mobile application in the decision process of whether an online- and offline-capable mobile application is a suitable choice, especially if developers are not familiar with the implementation of required components (e.g., local transaction manager, replication, synchronization/reintegration functionality) for such architectures. The approach is applicable for different types of mobile applications, for example, information and transaction systems, rather than for communication systems. Moreover, single-user systems and standalone applications without shared data structures can be developed. They mostly follow an offline-only architecture, which does not form the focus of this chapter.

The design approach can be used right at the beginning of a new development project, but it is also useful for the evaluation of existing, online-only mobile applications regarding their offline-capability. In such a scenario, application developers will initially model only parts (e.g., the core data structures and behavior) of the existing mobile application. Afterwards, model analysis and simulation clarify whether an online- and offline-capable version is advisable for the modeled application. The analytical app model can also be used for the generation of a software prototype.

Mobile application developers will profit from using the proposed model-driven design approach. Instead of analyzing the requirements for a mobile application and directly selecting an architecture, they can begin by creating an app model that is independent of certain architectural design decisions. These high-level app models are used for analysis and simulation, both of which are are tool-supported. Based on the results of these steps, mobile application developers can reason about an appropriate architecture for their mobile applications. The support of various architectures eases the development process, since mobile application developers can focus on business concerns instead of technical aspects.

REFERENCES

1. IBM DB2 Everyplace Installation and User's Guide. ftp://ftp.software.ibm.com/software/data/db2/everyplace/doc/enu/iug.pdf. (Accessed on August 13, 2018).
2. Adaptive Server® Anywhere 9.0.2—Adaptive Server Anywhere SQL User's Guide. http://infocenter.sybase.com/archive/topic/com.sybase.help.adaptive_server_anywhere_9.0.2/pdf/asa902/dbugen9.pdf, October 2004. (Accessed on July 23, 2018).

3. Adaptive Server® Anywhere 9.0.2—Introducing SQL Anywhere® Studio. http://infocenter. sybase.com/archive/topic/com.sybase.help.adaptive_server_anywhere_9.0.2/pdf/asa902/ dbfgen9.pdf, October 2004. (Accessed on July 23, 2018).

4. Adaptive Server® Anywhere 9.0.2—SQL Remote ™ User's Guide. http://infocenter.sybase.com/ archive/topic/com.sybase.help.adaptive_server_anywhere_9.0.2/pdf/asa902/dbsren9.pdf, October 2004. (Accessed on July 23, 2018).

5. Web Services Business Process Execution Language (WS-BPEL) Version 2.0. http://docs.oasis-open.org/wsbpel/2.0/OS/wsbpel-v2.0-OS.html, 2007. (Accessed on August 13, 2018).

6. Oracle® Database Lite Getting Started Guide. https://docs.oracle.com/cd/E1209501/doc.10303/ e12080.pdf, February 2010. (Accessed, on July 23, 2018).

7. Business Process Model And Notation (BPMN) Version 2.0. http://www.omg.org/spec/ BPMN/2.0, January 2011. (Accessed on August 13, 2018).

8. Study 'Der deutsche E-Commerce-Markt 2017'. https://www.e-commerce-magazin.de/studie-e-commerce-markt-deutschland-2017-multi-shop-und-marktplatz-strategien-ganz-vorne, 2014. (Accessed on August 13, 2018).

9. Amazon Europe Core S.a.r.l. http://www.amazon.de, 2015. (Accessed on August 13, 2018).

10. BookFit. https://www.bookfitapp.co.uk/, 2015. (Accessed on September 21, 2018).

11. GymJam. http://www.thegymjam.com/, 2015. (Accessed on August 13, 2018).

12. GymSync. http://www.gymsync.co.uk/, 2015. (Accessed on August 13, 2018).

13. Microsoft—Beginner's Guide to SQL Server Compact. https://msdn.microsoft.com/en-us/ data/ff687144, December 2015. (Accessed on August 13, 2018).

14. Otto (GmbH & Co KG). https://www.otto.de, 2015. (Accessed on August 13, 2018).

15. Teneo—version 2.0.0. https://wiki.eclipse.org/Teneo, 2015. (Accessed on August 13, 2018).

16. Zalando, S. E. https://www.zalando.de, 2015. (Accessed on August 13, 2018).

17. Android Sync Adapter. https://developer.android.com/training/sync-adapters/, 2017. (Accessed on August 13, 2018).

18. AndroidPIT International. https://www.androidpit.com/best-offline-android-apps, 2017. (Accessed on August 13, 2018).

19. Apple Pay. http://www.apple.com/apple-pay, 2017. (Accessed on August 13, 2018).

20. Google Developers—Firebase. https://firebase.google.com/, 2017. (Accessed on August 13, 2018).

21. Google Wallet. https://www.google.com/wallet/, 2017. (Accessed on August 13, 2018).

22. PayPal. https://www.paypal.com, 2017. (Accessed on August 13, 2018).

23. Smartface. https://www.smartface.io/, 2017. (Accessed on August 13, 2018).

24. A. Adya, B. Liskov, and P. O. Neil. Generalized isolation level definitions. In D. B. Lomet and G. Weikum, editors, *Proc. of the 16th International Conference on Data Engineering, San Diego, California, USA*. Institute of Electrical and Electronics Engineers (IEEE), Piscataway, NJ, USA, March 2000.

25. P. Bacon, R. Budde, K. Kautz, K. Kuhlenkamp, and H. Züllighoven. *Prototyping: An Approach to Evolutionary System Development*. Springer-Verlag, Berlin Heidelberg, Germany, 2012.

26. H. Berenson, P. Bernstein, J. Gray, J. Melton, E. O'Neil, and P. O'Neil. A critique of ANSI SQL isolation levels. In M. Carey and D. Schneider, editors, *Proc. of the 1995 ACM SIGMOD Int. Conf. on Management of Data, San Jose, CA, USA, May 22–25, 1995*. Association for Computing Machinery (ACM), New York, USA, 1995.

27. P. A. Bernstein, V. Hadzilacos, and N. Goodman. *Concurrency Control and Recovery in Database Systems*. Addison-Wesley, Boston, USA, 1987.

28. M. Book, V. Gruhn, M. Hülder, and C. Schäfer. A methodology for deriving the architectural implications of different degrees of mobility in information systems. In H. Fujita and M. Mejri, editors, *Proc. 4th Intl. Conference on Software Methodologies, Tools and Techniques (SoMeT 2005), September 28–30, 2005, Tokyo, Japan*. IOS Press, Amsterdam, Netherlands, 2005.

29. T. Bresnahan, J. Orsini, and P.-L. Yin. Platform choice by mobile app developers. *National Bureau of Economic Research (NBER) Working Paper*, 2014.

30. S. Ceri, F. Daniel, M. Matera, and F. M. Facca. Model-driven development of context-aware web applications. *ACM Transactions on Internet Technology (TOIT)*, 7(1):2, 2007.

31. P. K. Chrysanthis. Transaction processing in mobile computing environment. In B. Bhargava, editor, *Proceedings of the IEEE Workshop on Advances in Parallel and Distributed Systems*, volume 82, pages 77–82. Institute of Electrical and Electronics Engineers (IEEE), Piscataway, NJ, USA, October 1993.

32. J. Crinnion. *Evolutionary Systems Development: A Practical Guide to the Use of Prototyping within a Structured Systems Methodology*. Perseus Publishing, New York, NY, USA, 1992.

33. L. DeMichiel. Java persistence 2.0 specification. http://download.oracle.com/otndocs/jcp/persistence-2.0-fr-eval-oth-JSpec/, 2009. (Accessed on August 13, 2018).

34. R. A. Dirckze and L. Gruenwald. A pre-serialization transaction management technique for mobile multidatabases. *Mobile Networks and Applications*, 5(4):311–321, 2000.

35. M. H. Dunham, A. Helal, and S. Balakrishnan. A mobile transaction model that captures both the data and movement behavior. *Mobile Networks and Applications*, 2(2):149–162, 1997.

36. J. R. Escolar, C. G. Cachón, I. Marín, J. Vanderdonckt, and V. Motti. A model-based approach to generate connection-aware applications for the mobile web. *Romanian Journal of Human-Computer Interaction*, 7(2):117, 2014.

37. H. Garcia-Molina. Using semantic knowledge for transaction processing in a distributed database. *ACM Transactions on Database Systems (TODS)*, 8(2):186–213, 1983.

38. C. Gollmick. Konzept, Realisierung und Anwendung nutzer–definierter Replikation in mobilen Datenbanksystemen. *PhD thesis*, University of Jena, Germany, 2006.

39. J. Gray, P. Helland, P. O'Neil, and D. Shasha. The dangers of replication and a solution. In J. Widom, editor, *Proc. of the 1996 ACM SIGMOD Int. Conf. on Management of Data, Montreal, Quebec, Canada, June 4–6, 1996*. Association for Computing Machinery (ACM), New York, USA, 1996.

40. R. Hirsch, A. Coratella, M. Felder, and E. Rodriguez. A framework for analyzing mobile transaction models. *Journal of Database Management*, 12(3):36, 2001.

41. Z. Itani, H. Diab, and H. Artail. Efficient pull based replication and synchronization for mobile databases. In *International Conference on Pervasive Services 2005*, pages 401–404. Institute of Electrical and Electronics Engineers (IEEE), Piscataway, NJ, USA, 2005.

42. G. M. Kapitsaki, D. A. Kateros, G. N. Prezerakos, and I. S. Venieris. Model-driven development of composite context-aware web applications. *Information & Software Technology*, 51(8):1244–1260, 2009.

43. M. Keith and M. Schincariol. *Pro JPA 2*. Apress, New York, USA, 2013.

44. B. König-Ries. Challenges in mobile application development. *IT - Information Technology*, 51(2):69–71, 2009.

45. L. Kristensen, G. Taentzer, and S. Vaupel. Towards verification of connection-aware transactions models for mobile applications. In D. Moldt, L. Cabac, and H. Rölke, editors, *Petri Nets and Software Engineering. International Workshop, PNSE'17, Zaragoza, Spain, June 25–26, 2017. Proceedings, volume 1846 of CEUR Workshop Proceedings*, pages 227–228. CEUR-WS.org, 2017.

46. F. Laux and T. Lessner. Escrow serializability and reconciliation in mobile computing using semantic properties. *International Journal On Advances in Telecommunications*, 2(2), 2009.

47. J. Lee and K. Simpson. A high performance transaction processing algorithm for mobile computing. In H. Adeli, editor, *Proceedings Intelligent Information Systems. IIS'97*, pages 486–491. Institute of Electrical and Electronics Engineers (IEEE), Piscataway, NJ, USA, 1997.

48. J. Linwood and D. Minter. *Beginning Hibernate. Apresspod Series*. Apress, New York, NY, USA, 2010.

49. Q. Lu and M. Satyanaranyanan. Isolation-only transactions for mobile computing. *ACM SIGOPS Operating Systems Review*, 28(2):81–87, April 1994.

50. S. K. Madria and B. Bhargava. A transaction model to improve data availability in mobile computing. *Distributed and Parallel Databases*, 10(2):127–160, 2001.

51. B. Mutschler and G. Specht. *Mobile Datenbanksysteme: Architektur, Implementierung, Konzepte*. Xpert.Press, Springer, Berlin, Heidelberg, Germany, 2013.

52. P. E. O'Neil. The Escrow Transactional Method. *ACM Transactions on Database Systems (TODS)*, 11(4), 1986.

53. M. Ouzzani, B. Medjahed, and A. K. Elmagarmid. Correctness criteria beyond serializability. In L. Liu and M. T. Özsu, editors, *Encyclopedia of Database Systems*. Springer, New York, NY, USA, 2009.

54. P. K. Panda, S. Swain, and P. K. Pattnaik. Review of some transaction models used in mobile databases. *International Journal of Instrumentation, Control & Automation (IJICA)*, 1(1):99–104, 2011.

55. R. Pandya. *Mobile and Personal Communication Systems and Services. IEEE Series on Digital & Mobile Communication*. Wiley, Hoboken, NJ, USA, 2004.

56. E. Pitoura. A replication schema to support weak connectivity in mobile information systems. In R. Wagner and H. Thoma, editors, *Proceedings of the 7th International Conference on Database and Expert Systems Applications (DEXA '96)*, pages 510–520. Springer, London, 1996.

57. E. Pitoura and B. Bhargava. Maintaining consistency of data in mobile distributed environments. In *Proceedings of the 15th International Conference on Distributed Computing Systems*, pages 404–413. Institute of Electrical and Electronics Engineers (IEEE), 1995.

58. E. Pitoura and B. Bhargava. Data consistency in intermittently connected distributed systems. *IEEE Transactions on Knowledge and Data Engineering*, 11(6):896–915, November 1999.

59. E. Pitoura and G. Samaras. *Data management for mobile computing*. Springer Science & Business Media, New York, USA, 2012.

60. A. Puvvala, A. Dutta, R. Roy, and P. Seetharaman. Mobile application developers' platform choice model. In *49th Hawaii International Conference on System Sciences (HICSS)*. Institute of Electrical and Electronics Engineers (IEEE), Piscataway, NJ, USA, 2016.

61. M. Satyanarayanan. Fundamental challenges in mobile computing. In J. E. Burns and Y. Moses, editors, *Proc. Fifteenth Annual ACM Symposium on Principles of Distributed Computing, Philadelphia, Pennsylvania, USA, May 23–26, 1996*. Association for Computing Machinery (ACM), New York, USA, 1996.

62. E. Serral, P. Valderas, and V. Pelechano. Towards the model driven development of context-aware pervasive systems. *Pervasive and Mobile Computing*, 6(2):254–280, 2010.

63. P. Serrano-Alvarado, C. Roncancio, and M. Adiba. A survey of mobile transactions. *Distributed and Parallel Databases*, 16(2):193–230, 2004.

64. P. Serrano-Alvarado, C. L. Roncancio, and M. Adiba. Analyzing mobile transactions support for dbms. In A. M. Tjoa and R. R. Wagner, editors, *Proceedings 12th International Workshop on Database and Expert Systems Applications, 2001*, pages 595–600. Institute of Electrical and Electronics Engineers (IEEE), Piscataway, NJ, USA, 2001.

65. A. Sharma and V. Kansal. An analysis of mobile transaction methods and limitations in execution of m-commerce transaction. *International Journal of Computer Applications*, 43(21), 2012.

66. Q. Shi, D. Wen, and X. Yang. Study of logical transaction based data synchronization scheme for pervasive computing. In *World Automation Congress (WAC), 2012*. Institute of Electrical and Electronics Engineers (IEEE), New York, USA, 2012.

67. G. Taentzer and S. Vaupel. Model-driven development of mobile applications: Towards context-aware apps of high quality. In Lawrence Cabac, Lars Michael Kristensen, and Heiko Rölke, editors, *Proceedings of the International Workshop on Petri Nets and Software Engineering 2016, including the International Workshop on Biological Processes & Petri Nets 2016 co-located with the 37th International Conference on Application and Theory of Petri Nets and Concurrency Petri Nets 2016 and the 16th International Conference on Application of Concurrency to System Design ACSD 2016, Toruń, Poland, June 20–21, 2016, volume 1591 of CEUR Workshop Proceedings*, pages 17–29. CEUR-WS.org, 2016.

68. R. Tewari and P. Grillo. Data management for mobile computing on the internet. In R. Brice, C. J. Hwang, and B. W. Hwang, editors, *Proceedings of the 1995 ACM 23rd Annual Conference on Computer Science (CSC '95)*. Association for Computing Machinery (ACM), New York, USA, 1995.

69. H. Trætteberg. Using user interface models in design. In *Computer-Aided Design of User Interfaces III*, pages 131–142. Springer, New York, NY, USA, 2002.

70. S. Vaupel. A framework for model-driven development of mobile applications with context support. *PhD thesis*, Philipps-Universität Marburg—Fachbereich 12 Mathematik und Informatik, 2018.

71. S. Vaupel, D. Strüber, F. Rieger, and G. Taentzer. Agile bottom-up development of domain-specific ides for model-driven development. In Davide Di Ruscio, Juan de Lara, and Alfonso Pierantonio, editors, *Proceedings of the Workshop on Flexible Model Driven Engineering Co-Located with ACM/IEEE 18th International Conference on Model Driven Engineering Languages & Systems (MoDELS 2015), Ottawa, Canada, September 29, 2015, volume 1470 of CEUR Workshop Proceedings*. CEUR-WS.org, 2015.

72. S. Vaupel, G. Taentzer, R. Gerlach, and M. Guckert. Model-driven development of platform-independent mobile applications supporting role-based app variability. In J. Knoop and U. Zdun, editors, *Software Engineering 2016, Fachtagung des GI-Fachbereichs Softwaretechnik, 23–26. February 2016, Wien, Österreich, volume 252 of LNI*, pages 99–100. GI (German Association of Computer Science), Bonn, Germany, 2016.

73. S. Vaupel, G. Taentzer, R. Gerlach, and M. Guckert. Model-driven development of mobile applications for android and ios supporting role-based app variability. *Software and System Modeling*, 17(1):35–63, 2018.

74. S. Vaupel, G. Taentzer, J. P. Harries, R. Stroh, R. Gerlach, and M. Guckert. Model-driven development of mobile applications allowing role-driven variants. In J. Dingel, W. Schulte, I. Ramos, S. Abrahão, and E. Insfrán, editors, *Model-Driven Engineering Languages and Systems - 17th International Conference, MODELS 2014, Valencia, Spain, September 28–October 3, 2014. Proceedings, volume 8767 of Lecture Notes in Computer Science*, pages 1–17. Springer, New York, NY, USA,2014.

75. S. Vaupel, D. Wlochowitz, and G. Taentzer. A generic architecture supporting context-aware data and transaction management for mobile applications. In *Proceedings of the 3rd ACM International Conference on Mobile Software Engineering and Systems, MOBILESoft 2016, Austin, TX, USA, May 16–17, 2016*. Institute of Electrical and Electronics Engineers (IEEE), Piscataway, NJ, USA, 2016.

76. G. D. Walborn and P. K. Chrysanthis. Pro-motion: Management of mobile transactions. In B. Bryant, J. Carroll, D. Oppenheim, J. Hightower, and K. M. George, editors, *Proceedings of the 1997 ACM Symposium on Applied computing (SAC '97)*, pages 101–108. Association for Computing Machinery (ACM), New York, NY, USA, 1997.

77. G. Weikum and G. Vossen. *Transactional Information Systems: Theory, Algorithms, and the Practice of Concurrency Control and Recovery*. Series in Data Management Systems. Morgan Kaufmann, Burlington, MA, USA, 2002.

CHAPTER 3

Mobile App Testing

Tools, Frameworks, and Challenges

Silvia M. Ascate, Isabel K. Villanes, Kariny Oliveira,
Eduardo Noronha de Andrade Freitas, and Arilo Dias-Neto

CONTENTS

3.1 INTRODUCTION

Mobile applications are software systems that run on mobile devices and/or take in input contextual information [39]. Today, such applications are fundamental to performance of simple daily tasks, such as sending/receiving messages and playing games, but also in complex activities, such as banking operations or health care. For instance, mobile devices have been transformed into medical instruments that capture blood test results, medication information, blood glucose readings, and medical images [55]. Moreover, mobile applications continue to drive a large share of the information technology (IT) market. In March 2017, the number of mobile applications ("mobile apps" or just "apps"), available for download in Google's Play Store was around 2.8 million and in Apple's App Store was 2.2 million [52]. In June 2017, Apple announced that 180 billion mobile apps had been downloaded from its App Store [51].

Due to the growth in the number and complexity of mobile apps, quality has been a differential to obtaining advantages in a disputed market. Thus, the search for quality becomes a greater concern for such applications. However, these apps present some additional characteristics that are less commonly found in traditional software applications, some of which include supporting a wide range of mobile platforms, mobile devices, interaction with other applications, sensor handling, multiple input channels (e.g.: keyboard, voice, and gestures). Such characteristics have direct implications for software development and, consequently, for testing. Thus, it is important to pay more attention to these activities.

Mobile app testing is a collection of related activities with the aim of finding errors or uncovering issues in some dimensions of the mobile app quality, such as content, function, usability, navigability, performance, compatibility, and security [43]. Validating each dimension of the mobile app quality requires the design of testing activities with specific goals; that involves different types of testing. Mobile app testing is complex and high-cost and requires a lot of effort. When it is performed by a human tester, it becomes extremely repetitive and error-prone. Given this context, tester teams need to look into possibilities to increase the testing efficiency.

Automated testing is a strategy frequently used for ensuring mobile app quality. It involves multiple types of tests and frameworks. Knowing how to use all these tools is very important as it allows developer and tester to choose which tools and frameworks are appropriate for each type of testing.

This chapter provides an overview of mobile app testing, mainly regarding tools, frameworks, and services used to support mobile app testing in their diversity of levels. Also, a set of key issues and challenges surrounding diverse types of testing are broached.

3.2 MOBILE APP TESTING

The *term mobile app testing* refers to "testing activities for native and web applications on mobile devices using well-defined software testing methods and tools to ensure quality in functions, behaviors, performance, and quality of service, as well as features, such as mobility, usability, interoperability, connectivity, security, and privacy" [19]. This testing has been both manual and automated. It has often failed to involve the actual user in terms of user feedback, and analysis of app store reviews.

Mobile app testing can make use of the body of knowledge that exists for testing software in general, since nothing inherent to mobile devices invalidates already existing testing techniques and methods. However, this testing differs from the test approaches established in many IT organizations in some key areas:

Mobile system requirements: Compared to desktops, mobile systems bring with them a range of new requirements that companies need to take into account. Both Business and quality assurance (QA) experts need to work together with previous defined requirements to ensure that companies systematically and verifiably determine what properties any mobile software should have. In addition, the correct definition of requirements accelerates software development and testing.

Quality fingerprint: In principle, the quality criteria of mobile and traditional software are no different. The focus, however, shifts to functionality, security, performance, ease of use, reliability, portability, and maintainability. These criteria need to be weighted differently for mobile solutions, creating a different "quality fingerprint". Mobile computing is differentiated by four constraints: limited resources; security and vulnerability; the variability of performance and reliability; and ultimately finite power sources.

Multichannel testing: The basic principles of software QA and software testing remain the same with mobile systems. The essential aspect worth noting is that QA must ensure effective management of the many types of mobile software, e.g., different operating systems (OSs) such as iOS or Android, and also of different device classes such as smart phones or tablet PCs.

Testware: The traditional procedures of software development and QA continue to apply to mobile systems. However, considering the variety of OSs the same tool cannot be used on iOS and Android. Additionally, new testing tools need to consider the ability to test quality criteria such as security or efficiency.

Development and testing processes: When developing mobile systems, flexible process models are essential because mobile products change more quickly and frequently than traditional information technology. Iterative and incremental developmental models are recommended, as they test the system requirements and their implementation much more frequently than sequential procedures.

In a highly fragmented and competitive global market, the mobile development cycle is of short period. For the vendor's equanimity in the face of the overwhelming task of ensuring long-term success, the mobile app must be tested over different combinations of platforms, OSs, and networks before globally launching. In addition to this, similarly to functional testing, nonfunctional testing such as security testing and usability testing also play an important role. Effective test planning in mobile app testing helps to improve the mobile app quality. So knowledge about types of test and testing techniques that will be applied is essential to ensure the quality of the mobile app. Mobile app testing has four levels, which will be described in the next section.

3.3 TESTING LEVELS AND THEIR GOALS

Testing levels refer to the different testing scopes. For instance, unit testing focuses on small units of an application; integration testing considers the combined parts of an application to verify their combined functionality; system testing, the tests are executed on the whole application; and in acceptance testing, the tests are conducted to enable the customer to validate their requirements.

3.3.1 Unit Testing

Unit testing focuses on the verification of the smallest unit of the software component's or module's software design. It concentrates on the internal processing logic and data structures within the boundaries of a component/module [43]. In mobile apps, there are various components such as intents, content providers, services, and user interfaces (UIs). Therefore, it is important to define what units are to be tested. This level of testing is normally carried out by developers. According to Joorabchi et al. [26], the mobile community is still getting used to this level of testing and testing frameworks do not provide the same level of support for different platforms.

3.3.2 Integration Testing

Integration testing builds on unit testing to test sets of related application components in order to check how they work together and to identify failures due to their coupling [6]. The key component of such testing is the integration strategy, which defines the set of components to be integrated and tested as a whole. Some testing approaches consider mobile apps in isolation; however, interapplication communication via intents and content providers are also possible [39]. In this case, integration testing must contemplate external and internal information.

3.3.3 System Testing

System testing aims to discover defects in the underlying device and the real execution environment of the mobile app by testing the system as a whole [6]. This testing is generally performed in consecutive phases. In an initial phase, it can be executed on an emulator, because it enables observing the mobile app behavior in a limited context. However in the final phase, it is necessary to simulate a more realistic context. Therefore, it needs to be executed on the target mobile device.

3.3.4 Acceptance Testing

Acceptance testing is a formal testing with respect to user needs, requirements, and business processes conducted to determine whether or not a system satisfies the acceptance criteria [3]. It is challenging in several domains, including the mobile app domain, because it must capture the software's business expectation, which changes frequently and hence needs ongoing updating, refactoring, and maintenance [44].

3.4 TYPES OF TESTING

Mobile app testing is a collection of related activities with the aim of finding errors or uncovering issues in some dimensions of the mobile app quality such as content, function, navigability, performance, compatibility, usability, and security [43]. Validating each dimension of the mobile app quality requires the design of testing activities with specific goals. In other words, it involves different types of testing. The main types of testing in the mobile app context are functional testing, performance testing, security testing, compatibility testing, and usability testing.

A Motivational Example: To illustrate the concepts and theory of mobile app testing, we explore the Budget Watch application as our motivational example henceforward. This application allows users to manage personal budgets, expenses, and revenues. It is an open-source application from the F-Droid repository and belongs to the Finance category.

Figure 3.1 shows three screens of Budget Watch used to create a new budget. Main Activity presents the entrance of this application. Budget Activity shows a list of budgets during a configured period of time and an option to add a new budget. In Budget View Activity, the user enters the data of a new budget.

3.4.1 Functional Testing

Functional testing focuses on the functionality specifications of a component or a system, ignoring their internal mechanism. It concentrates solely on the outputs generated in response to selected inputs and execution conditions [1]. In mobile apps, functional testing validates service functions, mobile web application programming interfaces (APIs), external system behaviors, system-based intelligence, and UIs [19]. Both the mobile app and its environment should be specified, since the mobile app generally has limited resources, connectivity types, processing and data inputs, as well as other services and mobile apps that interact with the mobile device.

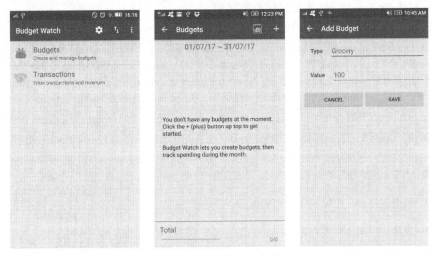

(a) MainActivity   (b) BudgetActivity   (c) BudgetViewActivity

FIGURE 3.1   Adding a new budget.

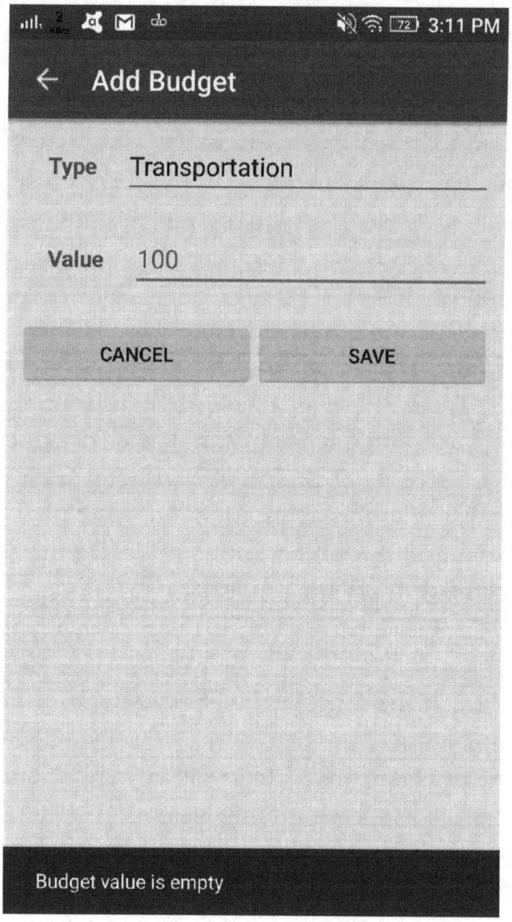

FIGURE 3.2   Creating a new budget with no value.

In the Budget Watch application, some functional tests must verify whether it allows for creating, updating, and deleting a budget. In creating a budget, the type and the value are mandatory. Thus, Figure 3.2 shows the Budget View Activity screen in which the budget value has not been entered. In this case, an error message is displayed. The test case to check whether the mobile app creates a new budget with no value is presented as follows:

1. Opening the mobile app.

2. Clicking on the "Budgets" option.

3. Clicking on the "+" button.

4. Typing "Transportation" in budget type.

5. Clicking on the "Save" button.

6. Checking whether the message "Budget value is empty" is displayed.

Some challenges in automated functional testing are related to the test case generation, the capture of context information, and isolating failures by mobile app layer. These challenges are detailed below.

1. Automatically generating functional testing is an open problem. Several issues are related to the inability to generate inputs and the creation of the appropriate test case scenarios [4,34,40].

2. Some context information may impact the mobile app behavior, such as physical location, multitasking, goal, and environmental changes. However, capturing them automatically is an issue [42].

3. Finding a mechanism to abstract context sources and derive test cases based on the most critical values [56].

4. Other issues are the difficulty of separating application-level failures from application frameworks or OS failures [39], deciding in which mobile devices to perform the testing [34,49], and testing for unexpected events.

3.4.2 User Interface Testing

User interface (UI) testing is one of the main types of functional testing. It is performed by interacting with the software under test via the UI [3]. It involves executing UI events from UI components and checking the correctness of states and behaviors, and verifying the data handling and control flows. UI testing is highly dependent on the UI, since if it changes the tests will have to be updated to take account of such changes. UI testing for mobile native apps embraces fat or smart mobile clients, rich media content and graphics, and gesture features. For mobile web apps, UI testing considers web-based thin mobile clients, downloadable mobile clients, and browser-based rich media and graphics support [19].

Figure 3.3 shows the Budget View Activity screen in which both budget type and value have been entered. After creating a new budget, it appears in the list in Budget Activity. The test case to check whether the mobile app creates a new budget is presented as follows:

1. Opening the mobile app.

2. Clicking on the "Budgets" option (see Figure 3.1).

3. Clicking on the "+" button.

4. Typing "Transportation" in budget type.

5. Clicking in "value".

6. Typing "300" as value for the budget.

7. Clicking on the "Save" button.

8. Checking whether the "Transportation" budget is created.

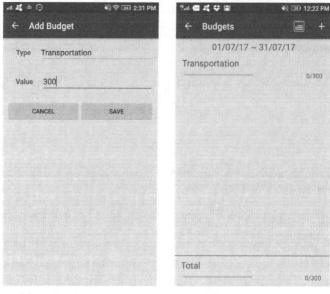

(a) BudgetViewActivity   (b) BudgetActivity

FIGURE 3.3   Creating a new budget.

Mobile app testing faces several challenges in automated UI testing:

1. Testing whether different devices provide an adequate rendering of data [39].

2. Testing whether native mobile apps are correctly displayed on different devices [39].

3. Generating sequences of unique events, in which each sequence represents a particular mobile app use and causes a change in the state of the mobile app [35].

4. Generating proper concrete values for UI data widgets that take user input [35].

5. In mobile web apps it is not trivial to determine when a page has finished loading [59].

6. Capturing some screen content and user-device interactions of mobile apps [49], for instance testing complex multitouch gestures [18].

3.4.3 Performance Testing

Performance testing is fundamental to mobile apps, mainly because most users expect mobile apps to start after four seconds or less, and they only tolerate a problem with a mobile app three times, after which they quickly abandon it [46]. This testing also aims to check the behavior of a system under a predetermined load [2]. In contrast to other types of testing, it must be supported by tools that simulate users with a given load. Performance testing refers to a series of tests designed to assess mobile app response time and reliability as demands on the server-side resource capacity increase [43]. In addition, mobile device resources (such as processing speed and memory size), network conditions, and mobile device operational mode must be considered when planning performance testing, since they affect the performance as application under test (AUT).

As with any testing, performance testing must have a defined objective, a structured approach, and parameters defined before being carried out. An objective of the Budget Watch application can be discovering performance based on different hardware/software configurations for starting such a mobile app and adding a new budget (see Figure 3.3) under varying network conditions. In Android applications, it verifies specified system performance goals, such as execution time, responsiveness, memory usage, power consumption, and network usage [6].

There are some challenges in automated performance testing:

1. Performance variations across devices are large, making it difficult to optimize performance while ensuring that the mobile app still works on all targeted devices [9].

2. Measuring the performance of mobile apps for various network bandwidths [9].

3. Discovering performance problems during early stages of development life, because in such stages emulators are used in place of real devices to simplify testing [28].

4. Determining whether a mobile app is suffering from performance bugs, since they gradually degrade the mobile app's performance.

5. Performance testing is affected by the lack of effective test oracles and test data generation techniques [31].

3.4.4 Security Testing

Movement of a device between networks with different security levels brings about several vulnerabilities for such a mobile device [39]. Security testing incorporates a series of tests designed to exploit vulnerabilities in the mobile app or its environment [43]. It aims to improve the mobile app's ability to protect against undesired accesses, improve its ability to protect system resources from improper use, and grant access to authorized users, services, and resources [6].

Testing activities for security testing check user authentication, device and session security, system/network penetration, peer-to-peer, mobile communications security, end-to-end transaction security, and user privacy [19]. For security testing (considering a mobile finance application), we can check whether this mobile app has access to personal information or whether this application is sending out personal information to unknown places or saving passwords.

Mobile security testing faces the following challenges [61]:

1. Signature-based malware detection techniques can be spoofed easily.

2. Penetration testing tools need to be improved for mobile devices.

3. A mobile app may hide its malicious activity when mobile security software is present.

4. There is a lack of effective approaches to detect data leakage when data is encrypted.

3.4.5 Compatibility Testing

Mobile apps exhibit different behavior patterns in different devices [23]. This is termed a fragmentation problem and is caused by the great variability of mobile platforms, OS versions, and hardware. Compatibility testing aims to find failures related to the variety of hardware, configuration profiles, and the application usage on nontargeted devices [6]. This testing does not demand the checking of the mobile app behavior against every possible device configuration, but only against configurations that the mobile app supports and any others relevant to users [6]. It is important to highlight that mobile apps can behave differently not only on mobile devices from different manufacturers but also on mobile devices from the same manufacturer [23,49].

In addition, selecting a set of mobile devices on which to execute compatibility testing is not a trivial task. Several works suggest maximizing diversity (feature) coverage and minimizing the number of devices [57]. The three main compatibility problems are platform compatibility, device feature compatibility, and native API compatibility [63]. A very common problem in compatibility testing is related to connection to the different types of networks, because mobile devices use different types of connections that are running at all times. The tester can perform a test, creating a Budget using three types of data connections (3G, 4G, and Wi-Fi), enable the Wi-Fi connection, perform some steps, and disconnect Wi-Fi. For this test, the Budget Watch application must save the data and not present any type of error when disconnecting the Wi-Fi connection. An example of this test case is presented below:

1. Connecting the device to a Wi-Fi network.

2. Opening the mobile app.

3. Clicking on the "Budgets" option.

4. Clicking on the "+" button.

5. Opening the top device menu and disabling the Wi-Fi connection.

6. Returning to the application.

7. Typing "Books" in budget type.

8. Typing "250" as value for the budget.

9. Clicking on the "Save" button.

Another example of this type of testing is related to different layouts displayed as calendar layouts or default buttons: for instance, in the mobile app used as an example, the difference between the layout of some default buttons as "Done", "Set", and "OK". The example is shown in Figure 3.4. These default buttons may change according to the platform API version and the brand of device.

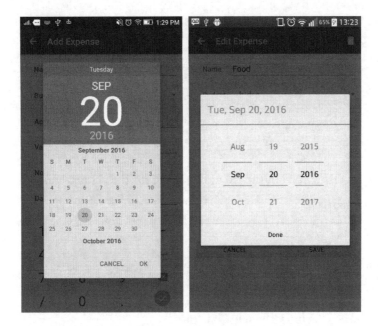

FIGURE 3.4   An example of the different layout of calendar and default buttons "OK" and "Done".

There are some challenges faced by compatibility testing as follows:

1. Compatibility testing needs to be done on numerous different devices, which takes a lot of time, effort, and money and is difficult to automate [49].

2. Automated scripting of tests needs to be abstracted away from the device to be of any real use; even more so if the tests are meant to be used on different devices [9,49].

3. Testing all the different device configurations on emulators would require more computing power than most developers have, while testing in the cloud is expensive [49].

3.4.6 Usability Testing

To provide a rich experience for their users, mobile apps usually include multiple input channels, and multimedia support [19]. Moreover, the smaller display, sensors, and location data play a dominant role in many mobile apps [60]. Usability testing helps to enhance the quality of the user experience on mobile devices. Its activities assess UI content and alerts, user operation flows and scenarios, media richness, and gesture interaction support [19]. Mobile apps can fail to accomplish their intended purpose simply because their mobile users are confused by the UI, and introduce erroneous data [12].

The focus should be the user experience, the ease and possibility of performing the intended actions. Based on this, using the Budget Watch application, the aim is to delete a budget. For evaluation, nonempirical techniques such as heuristic evaluation can be used. This is a diagnostic analysis technique in which a group of experts examines the interface and looks for problems that violate best practice. Gómez et al. [21] list some heuristics that can be applied in the usability evaluation of a mobile app. Flexibility and efficiency were chosen for illustration.

According to the chosen heuristics, the system needs to be easy for lay users, but flexible enough to become agile for advanced users. For evaluation, the exclusion of the Budget function was chosen. Simulating a lay user, there are two possible ways to exclude a budget:

First way:

1. Opening the mobile app.

2. Clicking on the "Budgets" option.

3. Pressing an element from the list and waiting for the context menu.

4. Clicking on "Delete".

Second way:

1. Opening the mobile app.

2. Clicking on the "Budgets" option.

3. Searching the "Exclude Element" button.

When performing any of the actions, the functionality is not found. An expert user could find the option using the following steps and the flow of Figure 3.5:

1. Opening the mobile app.

2. Clicking on the "Budget" option.

3. Pressing the list element until the context menu appears.

4. Clicking on "Edit".

5. Clicking on the "Edit" button.

6. Clicking on more options "…".

7. Clicking on the "Delete" button.

In this option, there are numerous steps to be performed, even for an expert user, which invalidates the heuristic of flexibility for that specific application.

There are some challenges faced by usability testing as follows:

1. Due to fragmentation problem it is challenging to enable natural interaction with the mobile device when performing usability testing [10].

2. Usability testing can suffer from the great variability of users' experience levels since such differences make it more difficult for a tester to assess the usability themselves, as well as to conduct a proper experiment [49].

3. The lack of definition of usability for mobile apps makes it difficult to measure [25,54].

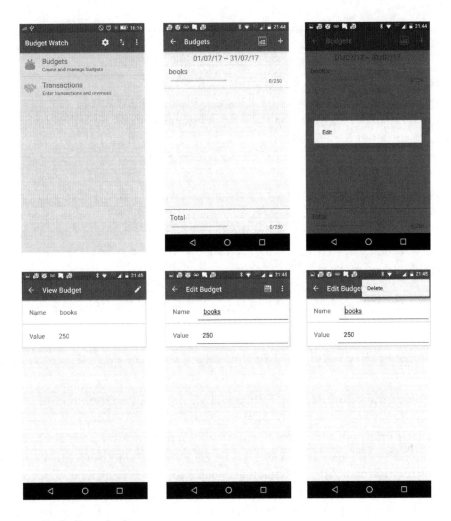

FIGURE 3.5   Excluding a budget.

There are no widespread principles about how to conduct usability field studies in the mobile app domain [42]. Moreover, it is not known how to prioritize different user-centered design evaluation criteria in mobile apps [25,54].

3.5 RELATED WORK

This section presents a review of studies regarding tools, frameworks, and challenges in mobile app testing.

To date, many papers have been published addressing different tools and frameworks to support mobile app testing. Some of them are described in the following. Gao et al. [19] provide an overview of the state-of-the-art tools and processes for mobile app testing. In their study, the authors discuss testing infrastructures such as emulation, device, cloud, and crowd based. The paper concludes with a discussion of challenges, issues, and needs for both native and web mobile apps. The main challenges in mobile app testing for the Android platform are analyzed by Amalfitano et al. [6].

In order to understand the test automation culture prevalent among developers, Kochhar et al. [29] investigate the current state of testing of mobile apps, the tools that are commonly used by developers, and the problems faced by them. The authors report that Android app developers use automated testing tools such as JUnit, MonkeyRunner, Robotium, and Robolectric. These tools are discussed in later sections of this chapter.

Linares-Vásquez et al. [33] analyze the developer's needs regarding practices and preferences in designing, automated testing, and quality metrics. They then present a perspective of automated mobile app testing with the purpose of reporting frameworks, tools, services, and challenges.

Developers often use blogs and Question & Answer (Q&A) sites in the search for solutions to their difficulties. Thus, Villanes et al. [58] investigate the main discussed topics on Android testing using repositories of the Stack Overflow Q&A site. The authors analyze testing tools, functional testing, and unit testing that were identified. They also analyze the evolution of interest in Android testing tools.

Muccini et al. [39] discuss the development and testing of the mobile app. Types of mobile applications, their peculiarities, and how they affect the research on mobile app testing are investigated.

3.6 MOBILE APP TESTING ENVIRONMENT AND ITS PARTICULARITIES

According to Starov and Vilkomir [50], mobile app testing has some similarities to website testing as both involve validation in many environments (smartphones and browsers, respectively). However, it also has specific testing characteristics such as different platforms, OSs, screen sizes, memory and processing capacity, models by vendors, and communication by different networks. Thus, testing all combinations of these characteristics could be unfeasible.

The construction of mobile app test environments still involves high costs and complexity levels. Setting up a mobile app test environment for multiple applications on each mobile platform for a range of devices is tedious, time-consuming, and expensive; frequent upgrades in both device and platform space only exacerbate this challenge [19]. Different mobile app test infrastructures are defined, such as emulation-based, device-based, and crowd-based testing, and cloud testing.

3.6.1 Emulation-Based Testing

Emulators are usually cheaper than other alternatives because there is no monetary investment and no testing laboratory and physical devices are necessary. However, previous knowledge is required for setup. The use of emulators is common because it comes with the software development kit (SDK) for development. The emulator simulates a device and displays it on a computer. Emulator users can configure the emulator according to their needs. Nevertheless, the use of emulators has distinct weaknesses and limitations in testing native device-based functions and behaviors.

3.6.2 Device-Based Testing

Devices are less common in a local test environment. The use of real devices implies that they have to be purchased. The tests are performed under real conditions (location, network), providing real insights into the functioning of the application. Performance of real devices is better than other virtual options. As a test environment, developers use their own smartphones (usually one device) to perform tests. However, this cannot cover all other mobile brands, or OS versions.

3.6.3 Crowd-Based Testing

Crowd testing is "a form of outsourcing which consists of carrying out testing tasks on a mass of internet users (the crowd)" [22]. In software crowd testing, applications are tested under real conditions, by real users, and with fast feedback. Crowd testing is software testing combined with the principle of crowd sourcing.

Crowd testing thus uses the collective intelligence of the global internet community to test websites, mobile apps, mobile games, and enterprise software. A number of crowd testing companies have been created in recent years (e.g., Applause, Mob4Hire, Pay4Bugs, 99Tests), reflecting the continued growth in software testing outsourcing.

3.6.4 Cloud-Based Testing

A new service model, known as testing as a service (TaaS) or cloud testing, has gained enormous popularity due to its scalability. The main idea is to create a cloud service that provides the ability to run tests on a variety of remote mobile devices (i.e., smartphones), which could be based on emulators or real devices.

This service promotes a cloud-based testing architecture to provide online testing services following a pay-per-use business model. In addition, it can be used to test environment or test tools, because all test activities and management are brought to the cloud [11]. TaaS combines two ideas: (1) offering software testing as a competitive, easily accessible web service, and (2) doing fully automated testing in the cloud, to harness vast, elastic resources toward making automated testing practical for real software [14].

3.7 MOBILE APP TESTING TOOLS, FRAMEWORKS, AND SERVICES

In this section, we provide a detailed analysis of popular mobile app testing frameworks, tools, and services. As frameworks and tools have different attributes from services, the analysis of the latter is performed separately.

3.7.1 Mobile Tools and Frameworks

Table 3.1 compares mobile tools and frameworks by type of testing, type of mobile apps, platforms, and some features. Tools and frameworks provide supporting for functional and UI testing include the following:

- *Robotium tool* [48] enables easy to write automatic UI tests. It allows the writing of function, system, and user acceptance test scenarios, spanning multiple Android activities.

- *Calabash* [13] an acceptance testing framework. It supports Cucumber, which expresses the mobile app behavior using natural language that can be understood by

TABLE 3.1 Mobile Testing Frameworks and Tools

Attributes	Robotium [48]	Calabash [13]	Monkey [37]	Appium [7]	Roboletric [47]	Monkeyrunner [38]	Espresso [17]	JUnit [27]	Mockito [36]	Ranorex [45]	UIAutomator [8]	MonkeyTalk [53]	Neoload [41]
Functional testing	Yes	Yes	No	Yes	No	Yes	Yes	No	No	Yes	Yes	Yes	No
UI testing	Yes	Yes	Yes	Yes	No	No	Yes	No	No	Yes	Yes	Yes	No
Unit testing	Yes	No	No	No	Yes	Yes	Yes	Yes	Yes	No	Yes	No	No
Security testing	No	No	No	No	No	No	No	No	No	No	No	No	No
Performance testing	No	No	No	No	No	No	No	No	No	No	No	No	Yes
Usability testing	No	No	No	No	No	No	No	No	No	No	No	No	No
Linux/Windows/Mac	Yes	Yes	Yes	Yes	Yes	Yes	Yes	Yes	Yes	Yes	Yes	Yes	L-W
Android OS	Yes	Yes	Yes	Yes	Yes	Yes	Yes	Yes	Yes	Yes	Yes	Yes	Yes
iOS	No	No	No	Yes	No	No	No	No	No	Yes	No	Yes	Yes
Windows OS	No	Yes	No	Yes	No	No	No	No	No	No	No	No	Yes
Mobile web applications	No	No	No	Yes	Yes	Yes	No	No	No	Yes	N/A	Yes	Yes
Mobile native applications	Yes	Yes	Yes	Yes	Yes	Yes	Yes	Yes	Yes	Yes	Yes	Yes	Yes
Mobile hybrid applications	Yes	Yes	Yes	Yes	Yes	Yes	Yes	No	No	Yes	N/A	N/A	Yes
Emulator/Device	Yes	Yes	Yes	Yes	Yes	Yes	Yes	Yes	Yes	Yes	Yes	Yes	Yes
Script language	Yes	Yes	No	Yes	Yes	Yes	Yes	Yes	Yes	Yes	Yes	Yes	Yes
Record and replay	No	Yes	No	No	Yes	Yes	No	No	No	Yes	No	Yes	Yes
Open source	Yes	Yes	Yes	Yes	Yes	No	Yes	Yes	Yes	No	Yes	Yes	No
Instrumentation	No	No	No	No	No	No	No	No	No	Yes	Yes	Yes	N/A
Natural language	No	Yes	No	No	No	No	No	No	No	No	No	No	No
Test reports	No	No	No	No	No	No	No	No	No	Yes	No	Yes	N/A

nontechnical people. It also supports all actions on the screen such as swipe, pinch, rotate, tap.

- *Monkey tool* [37] runs on an emulator or mobile device and generates pseudo-random streams of user events such as clicks, touches, or gestures, as well as a number of system-level events.

- *Appium* [7] is a test automation framework. Its library functions inside the framework make calls to the Appium server running in the background, which operates the connected device. It can be used in frameworks with various languages such as Java, Python, Ruby, and all others that Selenium web driver supports.

- *Monkeyrunner tool* [38] provides an API for writing programs that control an Android device or emulator from outside of Android code. It is designed to test the functionality of mobile apps and mobile devices and run unit test suites.

- *Espresso testing framework* [17] provides APIs for writing UI tests to simulate user interactions within a single target app. It provides automatic synchronization of test actions with the UI of the application under test (AUT).

- *Ranorex* [45] is a software testing tool that performs functional testing on desktop, web, or mobile apps. It supports many UI technologies that include Java, HTML, C#, Flex/Flash, Android, iOS, and Silverlight.

- *MonkeyTalk* [53] is a mobile app automation testing tool that automates real, functional, interactive tests. It can be used for simple smoke tests or for data-driven test suites.

- *UI Automator testing framework* [8] provides a set of APIs to build UI testing that performs interactions on user application and system applications. It enables: a viewer to inspect layout hierarchy, an API to retrieve state information, and APIs that support cross-app UI testing.

For performance testing, Neoload [41] is a load testing platform covering for cloud-ready applications, microservices architected applications, mobile and "internet of things" (IoT) applications, and enterprise-grade packaged application. Other types of testing, such as security and usability testing, have no support tools. These types of testing are usually supported by online services that will be detailed next subsection.

From Table 3.1 it can be seen that most applications are aimed at the Android platform; among the tools mentioned, only two have Windows Phone support (Appium and Neoload) and six have iOS support (Calabash, Appium, Ranorex, MonkeyTalk and Neoload). The wide adoption by the Android platform can be deduced as an impact of the growth and adoption of users to platforms, analyzing the current market and sales by platform is perceived a great advance in sales of Android devices, with iOS in second place, followed by other platforms.

Analyzing the tools (Table 3.1), most of the tools (10 of 13) support native and hybrid applications, accompanied by the growth of the use of the internet. It is increasingly common to find applications that use native elements in the mobile market and market trends have a great impact on such applications. Some characteristics that could influence in the choice of the testing tools are: the type of testing, the type of application (native, web or hybrid), the mobile platform and the testing environment. JUnit [27] is the most popular framework used to write unit tests. Mockito [36] and Roboletric [47] enable the writing of tests using the mocking approach. The former is used for making mocks of classes. It allows the mocking all of dependencies of a particular class. The latter contains many mocks of Android classes. It allows the tests to run on the java virtual machine (JVM) without booting up an instance of Android. Although Robotium, Espresso, and UIAutomator support functional UI testing, they can also be used to perform this testing.

Most tools and frameworks support script language except the Monkey tool. Using such techniques, the tester is required to manually write the test scripts that exploit the features offered by a test automation framework in order to automatically interact with the mobile app [5].

The record and replay technique reduces the effort of manually writing the test scripts since they require minimal coding at the beginning. Most tools and frameworks support record and replay. Although Robotium does not support such a technique, there is another tool—Robotium Recorder—that offers support for it, but it is not open source. Some framework and tool support partially captures multitouch gestures, for example, Espresso and MonkeyTalk. MonkeyRunner works with the screen coordinates, while other frameworks and tools that support UI testing allow identification of the screen components by ID and text.

Most tools and frameworks are open source, except for Ranorex and Neoload tools. Natural language is one of the formats preferred by Android developers for automatically generating test cases [33]. It is only supported by the Calabash framework.

3.7.2 Mobile Testing Services

The growth in the use of mobile apps raised the idea of mobile testing as a service (mobile TaaS). Mobile TaaS provides on-demand testing services for mobile apps and/or software as a service (SaaS) to support software validation and quality engineering processes by leveraging a cloud-based scalable mobile testing environment to assure predefined given quality of service (QoS) requirements and service-level-agreements (SLAs) [20].

These types of service became more popular due to their scalability and low cost. Xamarin Test Cloud,* AWS Device Farm,† Google Firebase,‡ SauceLabs,§ Testdroid¶ and Perfecto** are some companies that offer Mobile Testing Services. Of these, only Saucelabs uses emulators; the others use mobile devices for testing services. Only Google Firebase and Testdroid have the option to generate test cases. However, all of them are commercial services.

---

* https://www.xamarin.com
† https://aws.amazon.com/device-farm
‡ https://firebase.google.com
§ https://saucelabs.com
¶ https://bitbar.com/testing
** https://www.perfectomobile.com

3.8 DISCUSSION

3.8.1 Selection of Components for Unit Testing

During unit testing, determining which set of components to use in order to balance conflicting objectives such as cost of acquisition, development time, customer desirability, and expected revenue is a highly complex task [24]. Moreover, testing of all available components is computationally intractable [16]. SCOUT [16] treats the selection of components for unit testing with limited resources as an optimization problem, addressing two different objectives: maximizing benefits and minimizing testing cost.

3.8.2 Test Case Generation

Test case generation is the most costly task in software testing. This fact has increasingly been attracting researchers and practitioners to investigate ways to automate mobile app testing. Several new techniques for this purpose have emerged.

3.8.2.1 Generation of Test Data

Some techniques for mobile app testing are able both to automatically generate test cases and to execute them [5]. Such techniques use different strategies for generating test data such as random, model-based, or systematic exploration strategy [15]. Several tools support such techniques; for instance, Monkey and Dynodroid use random strategy, while A3e, Guiripper, and Puma use model-based strategies. Other tools such as Evodroid, ACTEve, and Sapienz implement a systematic strategy. Some existing mobile test generation techniques do not optimize the realism of tests, which may generate test cases with undesired properties, such as overlong sequences, too frequent events, or unnatural sequence patterns [32].

Otherwise, data augmentation refers to methods for constructing iterative optimization or sampling algorithms via the introduction of unobserved or latent data. This method was used to generate test data by exploiting the existing test data and was termed test data augmentation [62].

3.8.2.2 Generation of a Test Oracle

Test oracle is a mechanism to determine whether an application is executed correctly. A test oracle is usually determined by people who understand the software domain and it is therefore difficult to automate this step. In the mobile app domain, automatically generating a test oracle is extremely challenging due to specific characteristics of such applications, such as diversity of mobile devices and OSs, mobile connectivity, or limited resources.

Here, we present some tools that totally or partially support test oracle generation: the Barista tool [18] uses a record and replay technique that allows the creation of test scripts with built-in oracles that can run on multiple platforms. The QUANTUM tool takes the UI model of AUT, and interacts with it using a set of features (user actions such as rotation, killing and restarting) to generate test sequences that include test oracles. The SPAG-C tool [30] uses a record and replay technique to perform UI testing on Android devices using an integrated camera. Test oracles are generated from image verification with images captured from the camera.

REFERENCES

1. IEEE Std 610. *IEEE Standard Computer Dictionary: A Compilation of IEEE Standard Computer Glossaries.* IEEE, Piscataway, NJ, USA, 1991.

2. IEEE Std 829. *IEEE Standard for Software Test Documentation.* IEEE, Piscataway, NJ, USA, 1998.

3. ISTQB. *Standard Glossary of Terms Used in Software Testing.* International Software Testing Qualifications Board, Brussels, Belgium, 2015.

4. C.Q. Adamsen, G. Mezzetti, and A. Møller. Systematic execution of android test suites in adverse conditions. In *Proceedings of the 2015 International Symposium on Software Testing and Analysis, ISSTA 2015*, pages 83–93, ACM, New York, NY, USA, 2015.

5. D. Amalfitano, N. Amatucci, A.M. Memon, P. Tramontana, and A.R. Fasolino. A general framework for comparing automatic testing techniques of android mobile apps. *Journal of Systems and Software*, 125:322–343, 2017.

6. D. Amalfitano, A.R. Fasolino, P. Tramontana, and B. Robbins. Testing android mobile applications: Challenges, strategies, and approaches. *Advances in Computers*, 89:1–52, 2013.

7. Appium. Project website: http://appium.io/ [Accessed February 27, 2018].

8. UI Automator. Project website: https://developer.android.com/training/testing/ui-automator.html [Accessed February 27, 2018].

9. S. Baride and K. Dutta. A cloud based software testing paradigm for mobile applications. *SIGSOFT Software Engineering Notes*, 36(3):1–4, 2011.

10. J.M. Christian Bastien. Usability testing: a review of some methodological and technical aspects of the method. *International Journal of Medical Informatics*, 79(4):e18–e23, 2010. Human Factors Engineering for Healthcare Applications Special Issue.

11. K. Blokland, J. Mengerink, and M. Pol. *Testing Cloud Services: How to Test SaaS, PaaS, IaaS.* Rocky Nook Inc., Santa Barbara, CA, USA, 2013.

12. B. Bruegge and A.H. Dutoit. *Object-Oriented Software Engineering Using UML, Patterns, and Java.* Prentice Hall, Upper Saddle River, NJ, USA, 3rd edition, 2009.

13. Calabash. Project website: http://calaba.sh/ [Accessed February 27, 2018].

14. G. Candea, S. Bucur, and C. Zamfir. Automated Software Testing as a Service. In *SoCC '10 Proceedings of the 1st ACM Symposium on Cloud Computing*, Indianapolis, IN, USA. pages 155–160, ACM, New York, NY, USA, 2010.

15. S.R. Choudhary, A. Gorla, and A. Orso. Automated test input generation for android: Are we there yet? In *Proceedings of the 2015 30th IEEE/ACM International Conference on Automated Software Engineering (ASE), ASE '15*, pages 429–440, IEEE Computer Society, Washington, DC, USA, 2015.

16. E.N. de Andrade Freitas, C.G. Camilo-Junior, and A.M.R. Vincenzi. Scout: A multi-objective method to select components in designing unit testing. In *IEEE 27th International Symposium on Software Reliability Engineering (ISSRE)* pages 36–46, IEEE, Piscataway, NJ, USA, 2016.

17. Espresso. Project website: https://developer.android.com/training/testing/ui-testing/espresso-testing.html [Accessed on February 27, 2018].

18. M. Fazzini, E.N. De Andrade Freitas, S.R. Choudhary, and A. Orso. Barista: A technique for recording, encoding, and running platform independent android tests. In *IEEE International Conference on Software Testing, Verification and Validation (ICST)*, pages 149–160, IEEE, Piscataway, NJ, USA, 2017.

19. J. Gao, X. Bai, W.-T. Tsai, and T. Uehara. Mobile application testing: A tutorial. *Computer*, 47(2):46–55, 2014.

20. J. Gao, W.-T. Tsai, R. Paul, X. Bai, and T. Uehara. Mobile testing-as-a-service (mtaas) – infrastructures, issues, solutions and needs. In *IEEE International Symposium on High-Assurance Systems Engineering (HASE)*, page 158–167, IEEE, Piscataway, NJ, USA, 2014.

21. R. Yanez Gómez, D. C. Caballero, and J.-L. Sevillano. Heuristic evaluation on mobile Interfaces: A new checklist. *The Scientific World Journal*, Vol. 2014, Article ID 434326, 19 pages, 2014. http://dx.doi.org/10.1155/2014/434326.

22. F. Guaiani and H. Muccini. Crowd and laboratory testing, can they co-exist? An exploratory study. In *Proceedings 2nd International Workshop on CrowdSourcing in Software Engineering, CSI-SE 2015*, pages 32–37, IEEE, Piscataway, NJ, USA, 2015.

23. H.K. Ham and Y.B. Park. Designing knowledge base mobile application compatibility test system for android fragmentation. *International Journal of Software Engineering and Its Applications*, 8(1):303–314, 2014.

24. M. Harman, A. Skaliotis, K. Steinhöfel, and P. Baker. Search–based approaches to the component selection and prioritization problem. In *Proceedings of the 8th Annual Conference on Genetic and Evolutionary Computation, GECCO '06*, pages 1951–1952, ACM, New York, NY, USA, 2006.

25. A. Hussain, N.L. Hashim, N. Nordin, and H.M. Tahir. A metric-based evaluation model for applications on mobile phones. *Journal of Information and Communication Technology* 12:55–71, 2013.

26. M.E. Joorabchi, A. Mesbah, and P. Kruchten. Real challenges in mobile app development. In *ACM/IEEE International Symposium on Empirical Software Engineering and Measurement*, pages 15–24, ACM, New York, NY, USA, 2013.

27. JUnit. Project website: http://junit.org/junit5/ [Accessed February 27, 2018].

28. H. Kim, B. Choi, and S. Yoon. Performance testing based on test-driven development for mobile applications. In *Proceedings of the 3rd International Conference on Ubiquitous Information Management and Communication, ICUIMC '09*, pages 612–617, ACM, New York, NY, USA, 2009.

29. P.S. Kochhar, F. Thung, N. Nagappan, T. Zimmermann, and D. Lo. Understanding the test automation culture of app developers. In *IEEE 8th International Conference on Software Testing, Verification and Validation (ICST)*, pages 1–10, IEEE, Piscataway, NJ, USA, 2015.

30. Y.-D. Lin, J.F. Rojas, E.T.-H. Chu, and Y.-C. Lai. On the accuracy, efficiency, and reusability of automated test oracles for android devices. *IEEE Transactions on Software Engineering*, 40(10):957–970, 2014.

31. Y. Liu, C. Xu, and S.-C. Cheung. Characterizing and detecting performance bugs for smartphone applications. In *Proceedings of the 36th International Conference on Software Engineering, ICSE 2014*, pages 1013–1024, ACM, New York, NY, USA, 2014.

32. K. Mao. Towards realistic mobile test generation. In *Proceedings of the 2016 International Symposium on Software Testing and Analysis, ISSTA 2016*, ACM, New York, NY, USA, 2016.

33. K.M.M. Linares-Vásquez, C. Bernal-Cárdenas and D. Poshyvanyk. How do developers test android applications? In *Proceedings of the 33rd International Conference on Software Maintenance and Evolution, ICSME 2017*, Shanghai, China, 2017.

34. A. Méndez-Porras, C. Quesada-López, and M. Jenkins. Automated testing of mobile applications: A systematic map and review. In *Ibero-American Conference on Software Engineering*, Lima, Peru, 2015.

35. N. Mirzaei, H. Bagheri, R. Mahmood, and S. Malek. SIG-Droid: Automated system input generation for Android applications. In *IEEE 26th International Symposium onSoftware Reliability Engineering (ISSRE), 2015*, pages 461–471, IEEE, Piscataway, NJ, USA, 2015.

36. Mockito. Company website: http://site.mockito.org/ [Accessed February 19, 2018].

37. Monkey. Project website: https://developer.android.com/studio/test/monkey.html [Accessed March 5, 2018].

38. Monkeyrunner. Project website: https://developer.android.com/studio/test/monkeyrunner/index.html [Accessed February 27, 2018].

39. H. Muccini, A. Di Francesco, and P. Esposito. Software testing of mobile applications: Challenges and future research directions. In *7th International Workshop on Automation of Software Test (AST)*, pages 29–35, June 2012.

40. M. Nagappan and E. Shihab. Future trends in software engineering research for mobile apps. In *IEEE 23rd International Conference on Software Analysis, Evolution, and Reengineering (SANER)*, volume 5, pages 21–32, IEEE, Piscataway, NJ, USA, 2016.

41. Neoload. Company website: http://www.neotys.com/neoload/overview [Accessed February 19, 2018].

42. A. Oulasvirta. Rethinking experimental designs for field evaluations. *IEEE Pervasive Computing*, 11(4):60–67, 2012.

43. R. Pressman and B. Maxim. *Software Engineering: A Practitioner's Approach*. McGraw-Hill, New York, NY, USA, 8th edition, 2015.

44. M. Rahman and J. Gao. A reusable automated acceptance testing architecture for microservices in behavior-driven development. In *IEEE Symposium on Service-Oriented System Engineering*, pages 321–325, IEEE, Piscataway, NJ, USA, 2015.

45. Ranorex. Company website: https://www.ranorex.com/ [Accessed February 27, 2018].

46. Dimensional Research. Failing to meet mobile app user expectations: A mobile app user survey. website: http://www.dimensionalresearch.com, 2015 [Accessed March 5, 2018].

47. Robolectric. Project website: http://robolectric.org/ [Accessed February 27, 2018].

48. Robotium. Project website: https://github.com/robotiumtech/robotium [Accessed February 19, 2018].

49. T. Samuel and D. Pfahl. *Problems and Solutions in Mobile Application Testing*, pages 249–267. Springer International Publishing, Cham, Switzerland, 2016.

50. O. Starov and S. Vilkomir. Integrated TaaS platform for mobile development: Architecture solutions. In *8th International Workshop on Automation of Software Test (AST)*, pages 1–7, IEEE, Piscataway, NJ, USA, 2013.

51. Statista. Cumulative number of apps downloaded from the Apple App Store from July 2008 to June 2017 (in billions): https://www.statista.com/statistics/263794/number-of-downloads-from-the-apple-app-store/ [Accessed March 5, 2018].

52. Statista. Number of apps available in leading app stores as of March 2017: https://www.statista.com/statistics/276623/number-of-apps-available-in-leading-app-stores/ [Accessed March 05, 2018].

53. Monkey Talk. Project Website: https://www.perfomatix.com/blog/mobile-app-automation-testing/ [Accessed February 19, 2018].

54. L. Tang, Z. Yu, X. Zhou, H. Wang, and C. Becker. Supporting rapid design and evaluation of pervasive applications: Challenges and solutions. *Personal Ubiquitous Computing*, 15(3):253–269, 2011.

55. TechTarget. Mobile devices, apps and the patient health management revolution: http://searchhealthit.techtarget.com/feature/Mobile-devices-apps-and-the-patient-health-management-revolution [Accessed March 05, 2018].

56. V. Vieira, K. Holl, and M. Hassel. A context simulator as testing support for mobile apps. In *Proceedings of the 30th Annual ACM Symposium on Applied Computing, SAC '15*, pages 535–541, ACM, New York, NY, USA, 2015.

57. S. Vilkomir. Multi-device coverage testing of mobile applications. *Software Quality Journal*, 26(2): 197–215, 2018.

58. I.K. Villanes, S.M. Ascate, J. Gomes, and A.C. Dias-Neto. What are software engineers asking about android testing on stack overflow? In *Proceedings of the 31st Brazilian Symposium on Software Engineering, SBES '17*, ACM, New York, NY, USA, 2017.

59. X.S. Wang, A. Balasubramanian, A. Krishnamurthy, and D. Wetherall. Demystifying page load performance with wprof. In *Presented as part of the 10th USENIX Symposium on Networked Systems Design and Implementation (NSDI 13)*, pages 473–485, Lombard, IL, USENIX, Berkeley, CA, USA, 2013.

60. A.I. Wasserman. Software engineering issues for mobile application development. In *Proceedings of the FSE/SDP Workshop on Future of Software Engineering Research, FoSER '10*, pages 397–400, ACM, New York, NY, USA, 2010.

61. Y.A.Y. Wang. Mobile security testing approaches and challenges. In *First Conference On Mobile And Secure Services*, February 19–21, 2015, Gainesville, Florida, February 2015.
62. S. Yoo and M. Harman. *Test data augmentation: generating new test data from existing test data*. Centre for Research on Evolution, Search Testing (CREST) King's College London, Technical Report: TR-08-04, 2008.
63. T. Zhang, J. Gao, J. Cheng, and T. Uehara. Compatibility testing service for mobile applications. In *IEEE Symposium on Service-Oriented System Engineering*, pages 179–186, IEEE, Piscataway, NJ, USA, 2015.

CHAPTER 4

City Tour App

*A Step-by-Step Development
of an e-Commerce Platform*

Han Jing, Jonathan Loo, Michael Chai, and Junaid Arshad

CONTENTS

4.1 INTRODUCTION

The City Tour app is inspired by the Uber app, which provides a successful e-commerce platform that caters to the taxi trade whereby any registered private driver can be a taxi service provider to meet the rider's requirements. This app focuses on implementing an Android application to fulfil requirements within this context and adds more functions for tourism. The app allows a user to publish a request consisting a tour plan and wait for a local driver who can offer a ride as well as make suggestions for this trip. Optimizing the route and optimal price calculation are fundamental to achieving this. The app contributes to travel intelligence and the sharing economy, providing business opportunities for both tourism and the taxi industries.

4.1.1 Motivation

The quick, reliable, and cheap chauffeur car service is one of the most popular services and it provides many potential business opportunities. However, contemporary services do not take into account specific requirements of tourists. For instance, as reported by Sri Lanka's *DailyFT* [23]), 31.5 million tourists visited the city of London in 2015, a number expected to grow significantly. This huge number of tourists visiting represents an extraordinary user-base and potential business opportunities, and therefore an app targeted at these users has the potential to benefit many tourists and transform the rules of the taxi and tourism industries.

Mirroring the historical local guide, this project aims to design and implement an e-commerce platform that enables users to travel in comfort, helping them to generate an optimal route, a reasonable price, and an option to do a trade by offering a personalized tour service. Furthermore, we envisage that the underlying algorithm has the potential to be enhanced with machine learning and larger datasets to achieve *travel intelligence*. Travel intelligence has a focus on the creation of value from data and it has already become immersed in modern life. For example, Airbnb landlords can use a dynamic pricing model to predict the probability of renting out their house at a certain price. Similarly, Uber analyzes traffic data to predict, for example, whether the user will be late for a lunch appointment.

4.1.2 Technical Context

The application is developed within Android Studio using Java programming language. This app is compatible with Android versions 4.4 (KitKat, API 19) and up, but it is recommended for Android versions 5.0 (Lollipop, API 21) and above. The app also uses AMAP API to provide the maps for its users. Choice of this API (application programming interface) is influenced by its ability to offer strong map re-development capabilities, a comprehensive map support, including off-line map, data support, and a variety of interactive map modes with the aim of meeting the needs of different scenes on the map.

4.2 REQUIREMENT ELICITATION

4.2.1 Requirement Elicitation

As discussed, the app is a combination of taxi app and a tourist app—it not only offers taxi-like functions but also provides the tourists with a plan for traveling around the city. Consequently, observations were made on existing apps related to this project including Uber, DiDi Taxi, Airpano, and Ctrip apps. The common functions of those apps were observed and the extensional parts were understood. Then interviews and questionnaires were conducted with users to conceptualize their requirements as the "baseline functions" and "interested functions" need to be clarified for the users since the specification only takes some examples but does not limit the specific functions.

4.2.1.1 Observation

To design an appropriate interface for the City Tour app and understand the users' requirements, four popular taxi and tourism apps were studied and compared. Those apps are fully integrated solutions that include functions necessary either to take a ride or to plan a trip. But none of them provides a full solution to this project; thus the observation aimed to identify the parts so as to be able to extend beyond them. Table 4.1 shows a comparison among those apps.

4.2.1.2 Interview

Personal interviews were employed as part of the requirement elicitation methodology. Within this context, 10 people were directly contacted in face-to face interviews to investigate a customer-centric view on overall design. The interviewees were either interested in traveling or frequent users of taxi apps. One-to-one interviews about the trip app, taxi app and their thoughts and suggestions were important for this project. The interview

TABLE 4.1 Comparison Among Tour/Taxi Apps

Name	Type	Main Features	Operational Domain to Extend
Uber	Taxi app	Develops, markets and operates the Uber car scheme. Uber drivers use their own cars or rent a car to offer a ride to users.	Does not offer a plan for tourists new to a city, and the user can only select a specific destination and a specific starting point.
DiDi Taxi	Taxi app	Adds more types of cars than Uber. A particular type is "hitchhiking" which is more related to this project.	Lacks tourism features but offers a way to combine two roles in an app that is implemented in our City Tour app.
Airpano	Tourism app	Implements all the functions for high comfort with high-quality introduction and pictures about destinations.	More functions can be identified relating to tourism by referencing this app, such food and place introductions.
Ctrip	Tourism app	Provides tourists with functions relating to hotel and ticket ordering, and has detailed information about each destination.	Top 10 destinations in the city are identified and this project adds a function termed "personal requests" by which users can click destinations from what the app lists.

questions mainly focused on what kind of service the users want to receive from such an application. It was concluded from the interviews that apart from the reliability and basic functions of the app, users most wanted an app that could guide them in a new city with an easy-to-use user interface. This requires that the app's activities are well organized without any complexity in operation. The simplicity of the app, to some extent, contributes toward the reliability. Also, more "clicks" should be made instead of "choices" in this project, as according to the interviews it is more convenient for users to make a decision instead of reading introductions of each choice in detail.

4.2.1.3 Questionnaire

Another instrument used to collect user input in this project was questionnaires. The questionnaire was designed to collect possible requirements regarding functionality and user interface design. Overall, the questionnaire consisted of 15 questions and 92 feedbacks were received. As the outcome of this exercise the following points were concluded: First, basic functions such as *publish a request, get a ride offer, manage the routes, calculate the corresponding price*, and *offer tourism information* are necessary. Second, when the driver accepts the offer, the status in the user side should change. Third, the user needs to be able to select the mode of publishing a ride—either by entering the destinations they know or by clicking a destination that is popular in the city. In addition, more than half of the users (72.2%) think that a bright color will be more suitable for this app as it makes people happy and this is important during traveling. Finally, flexible software interfaces with functions compatible with different operations are essential.

4.2.2 Functional Requirements

The main functions of the app are presented below.

1. Both the rider and the driver can publish a request and see available requests, and the driver can offer a ride with a corresponding price, after which the route displays on the main page.

2. The rider can select a mode to publish the request—either selecting the destinations from 10 places or searching the destination and entering it.

3. The plan displays after the rider selects the destinations, starting time, end time, and the number of people.

4. When the driver accepts a ride, the status in the rider side will change accordingly; and before that the rider can cancel the ride.

5. The rider can look for local entertainment, weather information, and places nearby.

4.2.3 Nonfunctional Requirements

- The application should be developed into one app; to build the application development environment, Android Studio and SDK (software development kit) tools needs to be downloaded.

- The application should be developed with a database, for example, JDBC (Java Database Connectivity) to connect the server and database and MySQL database.

- *Usability*: The user interface should be easy for new users to use even without a user manual.

- *Flexibility*: The app is intended to be able to adapt to continuously changing requirements.

- *Reliability*: The app should have error messages when users have error operations.

- *Privacy*: The system cannot disclose the personal information of the driver or the rider.

4.2.4 Functionality to be Realized

The methods mentioned above help the developer to analyze the requirements and understand the functionality to be realized. Those methods translate the original ideas into objectives, functional requirements, nonfunctional requirements, and business rules. The constraints of the app are also elucidated by application of these methods.

According to the requirements elicited, there are several functions to be realized. First, the user can view the map and choose the starting point and destinations, after which a ride request may be sent. The request consists of a calculated price and an optimized route. When a driver accepts a user's request and decides to offer the trip, he or she could click the "start" button. Then a driving route and the price of the trip will display on the map. Meanwhile, the status of the request will update so that the user is able to note that someone will offer a ride. Figure 4.1 shows the flowchart of the interactions of a user with the proposed application, including the main functionalities of the application.

4.3 DESIGN AND IMPLEMENTATION

4.3.1 Design

The background research and requirement questionnaire mentioned above enabled us to understand user requirements for the City Tour application. These instruments focused on identifying user expectations while traveling around a new city and can help to make the application more user-friendly and useful. This information was fundamental to developing City Tour app with the functionalities and UI design and led to development of a prototype of this application and the links between different activities. Meanwhile, based on the requirements and functionalities, the business logic and algorithms for price and route calculation were designed. Furthermore, the user interface is optimized continuously in order to make the use of this app consistent with the business logic.

Consequently, the architecture of the application was considered, including layering the application, handling internet connection, encapsulating the data and activities.

4.3.1.1 User Interface Design

Figure 4.2 shows the steps of user interface design. After collecting customer requirements by online questionnaire, data analytics on customer requirements was performed in terms

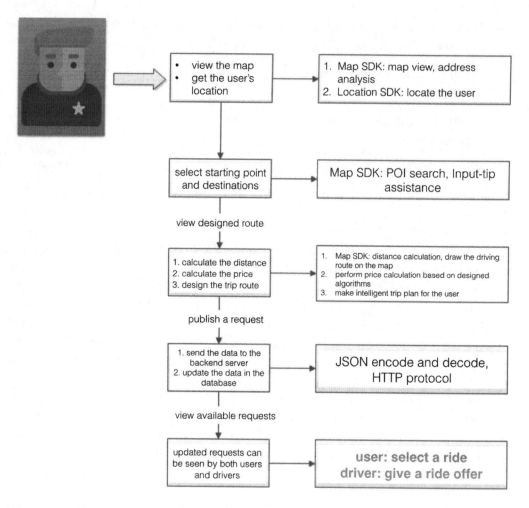

FIGURE 4.1   User interaction with the application.

of both ease of customer requirements and technical implementation, thus optimizing the overall design. A major change is to transform from two separate applications—one for trip drivers and the other for tourists, with separate login and business logic—to two roles in a single app. Each login requires a judgment, and subsequently different interfaces will display. Although this is a challenge for both the performance of the app and mobile phone resources, considering the large overlapping functions of both roles the two sides require almost the same functions. The driver can be a tourist in a new city and act as a tour guide in his or her own city, just as with Airbnb the lessor and the occupants need a flexible exchange of both roles. Thus we transformed this application into a single app and the two roles can be switched. Figure 4.3 shows the view of role selection in this app.

questionnaire ➡ functionality ➡ prototype ➡ layouts

FIGURE 4.2   Steps of user interface design.

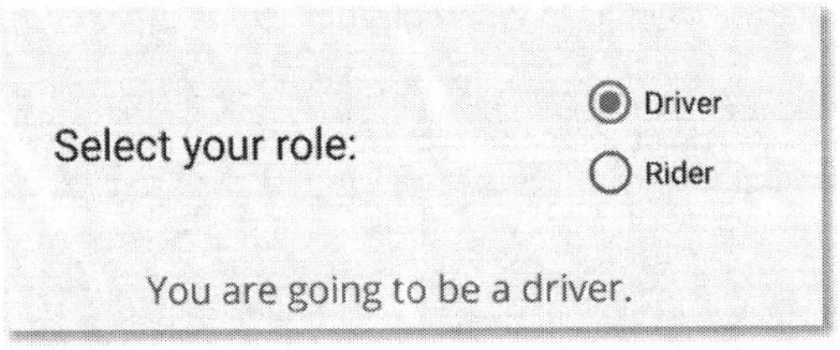

FIGURE 4.3   View of role selection.

As for the basic features to be implemented, the application behaves as a standard Android application and has Uber-like features similar to those of a taxi service. A user can be in one of two roles:

1. *Rider*: A rider can request a ride while looking at current location (which is updated very frequently) on a map. Once the rider clicks the "Request Ride" button, the request is stored on the online database. The rider also has the option to cancel the request, which will in turn remove it from the database. Any drivers can see the request from riders.

2. *Driver*: A driver can view available requests from other riders displayed in an Android ListView, which is a vertical table with clickable elements. The requests ordered on the basis of time. Once drivers click on a request, they can accept it or look for another one. Accepting the request brings up a navigation view of the whole route as well as the price view of this ride.

The user requirements are then conceptualized, and the user interface design prototype and optimization is made by the Prototyping on Paper (POP) app, which helps to generate an interactive user interface. Table 4.2 shows the steps of user interface design and optimization.

The resources used to improve the UI design include Material Design of Google, the Dribble website, and the open resource widgets in Github. The whole process begins from designing the home page, making choices among different kinds of menus and buttons, and selecting the mainstream color to coding the component layouts. The final part was to code several layout designs using XML and make links between different pages using Java, whose main steps are selecting views, positioning views, and styling views.

4.3.1.2 Business Logic Design

The design of business logic requires consideration of the sequence and data usage of the app, such as accounts and trip rides that determine how data can be created, saved, and altered. This leads to considerations of lower-level details of this app, such as displaying the user interface, managing the database, and connecting different parts of the application.

In this project, the e-commerce app allows visitors to add requests to a plan lists, specify a traveling time and destinations, and supply price information. Figure 4.4 shows the main business logic.

The sequence of events that happens during this app is that first the user needs to register or log in using the available account and then uses this app to publish a request or take an

TABLE 4.2   Steps of User Interface Design and Optimization

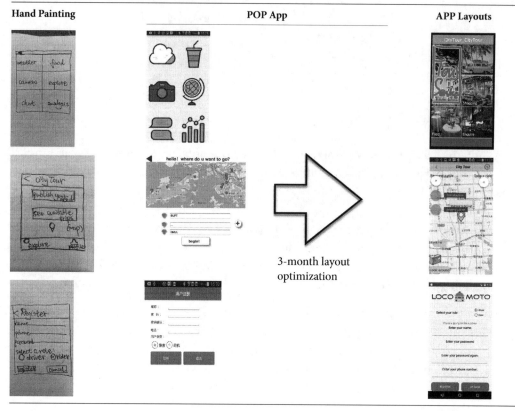

available ride. To publish the need, there are two modes: the user could generate his or her own plan (normal mode) or use this app to generate a trip plan (personal mode). In normal mode the logic is the same as a classic taxi app. In personal mode the app asks for the starting point, destinations, time, and the number of people, then the next page will show the plan generated by the app, consisting of the order of destinations, visiting time, driving time, and time of arrival at each destination; the last page asks whether the user accept this plan and, based on the user's answer, it will either show congratulations or redirect to the previous page.

To summarize, the business logic is as follows:

1. The user is allowed to create a list of places (or point of interest) with priorities; the City Tour app can generate an optimal route and calculate the cost of travel based on distance, time, labor, etc.

2. Publish this as a request to the registered private drivers/tour operators as a potential trade deal.

3. The registered private drivers/tour operators can generate a suggested tour with a fairly calculated cost.

4. Advertise to users.

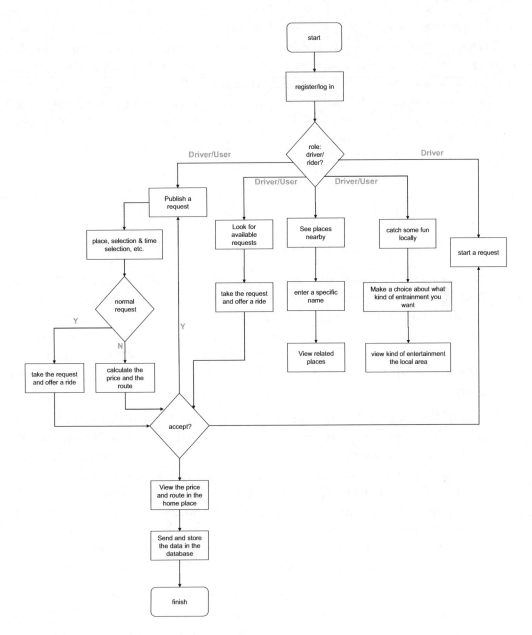

FIGURE 4.4   Business logic for the application.

The business rules of the application are as follows:

- Increasing the number of people and the distance between destinations from the request publishing page increments the price for that trip.

- Specific formats of the rider's destinations, mobile phone number, the number of traveling days and other personal information must be followed.

- A specific internet protocol and a format of data for communicating with the back-end server.

4.3.1.3 Algorithm Design

In this app, there are two parts of algorithm design, which are price calculation algorithm design and route calculation algorithm design. They are used for automatically generating the plan route with corresponding price for the user. In this section, the principle of the algorithm is eliminated. This section aims to guide developers to use their algorithm into the app instead of the way to design a complex algorithm. Therefore, this section selects the price calculation algorithm to introduce the basic knowledge, which is easier to understand.

4.3.1.3.1 Price Calculation Algorithm Design   In the design of this algorithm, four factors were considered and the price is calculated based on Equation (4.1):

$$Price = f(\text{distance, traffic, day, time}) \tag{4.1}$$

1. *Ride time (distance).* As well as the distance, the ride time (duration) should also be a factor in calculating the final price. Initially the price is automatically calculated based on GPS positioning and street data; then, with some alterations to the traditional algorithms, it will be adjusted according to the time commonly used to a specific distance. This is the most critical difference with the ordinary taxi service— the passengers pay a price that is based on the time spent rather than only on the distance.

2. *Traffic.* The traffic conditions can affect the length of time taken; that is, the price will be adjusted according to the level of traffic. When the traffic is busy, a journey of the same distance will take longer time. This pricing strategy will encourage drivers to offer ride to passengers in peak time, and at times of low demand the drivers can stay at home. Uber patented this pricing strategy based on a large dataset, also known as "peak pricing strategy." It is also termed "dynamic pricing"; this is very similar to the on-demand pricing used by chain hotels and airlines, but Uber's implementation is more complicated, not simply raising prices on weekends or public holidays but using forecast models to evaluate real-time requirements.

3. *Particular time.* When the trip request is started at peak time in the morning or in the afternoon, the price will be higher than normal. For example, at 8 a.m. there is high volume of traffic and the driving time will be much higher than it is one or two hours later. Increasing the price is reasonable because the ride time will be increased and thus the fuel cost will increase.

4. *Special days.* The demand will vary on different days. For example, during Christmas, the trip demand will be much higher than usual. Similarly, imagine the end of a concert late in the night when buses are no longer in service and at a time when car services are in short supply. This will trigger the price to rise, which is good as the price drives the balance of supply and demand. Thus, when the user's waiting time appears relatively steep, it will engage the price increase algorithm. Most of these cases are predictable, such as around holidays or certain annual concerts and events.

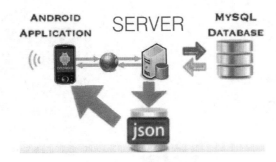

FIGURE 4.5   Overview of the system.

4.3.2 Implementation

The technical components are implemented as APP–Server–Database as presented in Figure 4.5.

The android app sends an HTTP client request and receives a response from the server, and the server connects to the MySQL database, which will return to users whatever they request and display the results.

We use Android Studio as the integrated development environment (IDE) for Android application development, MyEclipse to develop the back-end server, and MySQL as the database.

4.3.2.1 Front-End Application

4.3.2.1.1 APP: Android Studio   Android Studio, as the official IDE for the Android platform, is used throughout this project. In this part, the implementation of this app will be explained together with the Java files related to a specific function. Although it will be explained here how the front-end application interacts with the back-end system, the detailed implementation of the back-end system will be shown in the next section.

In the following, the implementation of the main functions of this app will be described with Java files in folders along with the main methods in detail. The order will be the same as that in using the app.

4.3.2.1.2 Login & Logout/Home Page   Figure 4.6 shows screenshots of login–logout and home pages. Before entering the home page, the user needs to either login in or register. After the user enters the required information, a validation function named **validate()** will be used in logging in and registering. If all the fields are filled both correctly and completely, then the data will be sent in JSON (JavaScript Object Notation) format to the back-end database to be stored. For instance, if the user uses an invalid phone number or leaves a blank, then an alert dialog will display. Next time, a registered user could simply log in with correct information and enter to the home page. There are three activities responsible for this:

- ActivityLogin

- ActivityRegister

- MainActivity

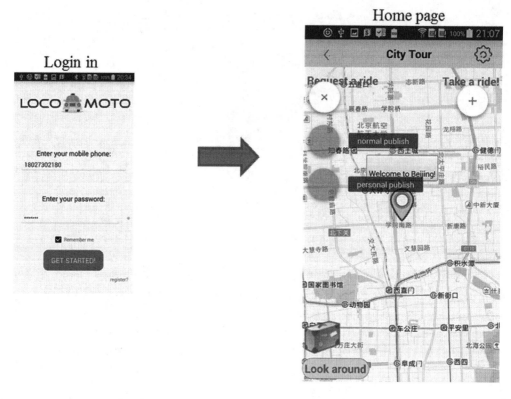

FIGURE 4.6   Login & logout/home page.

ActivityLogin and **ActivityRegister** are activities handling login in and registration, respectively. The user will use the phone number as a identifier and log in in using correct password. The underlying logic is that the application will send the register data of users to server, and the server will handle the request and store the data in the **MySQL** database. When the user logs in, the server will query the database whether the user exists and send the result to the application. To parse the data into **JSON** format, we used a Google library called "**GSON**," which we will introduce in the next section.

A function named "remember me" is also added in the app, which is implemented by using **SharedPreferences** class, a lightweight class for storing simple data including int and float number. This is used to store the password and phone number of a user so that the next time the user wants to log in, he or she does not need to enter the information repeatedly.

MainActivity is responsible for the home page. It offers multiple functions to users with different roles. It contains a map and some buttons at the head that the user can select to publish a request, take a ride, or look for entertainment nearby. In "driver" role, if a specific ride is selected, the route will be displayed on the map.

To make the map view available, **AMAP API** is used, which is introduced in the earlier part of the chapter. The implementation of this part will be described only briefly since it is also shown in detail on the official website [10]. To begin with, it is necessary to add *.jar* and *.so* files offered by **AMAP**. To use the API, the developer should register and obtain the development key from the official website, and then **AndroidManifest.xml** has to be deployed

FIGURE 4.7   Publish a request.

in the project to set the permissions and the key. In the home page, different view will be shown in different roles ("rider" or "driver"). This is achieved by a variable called **type** that is stored in the database. On identifying the number of the type, the app will display different views.

4.3.2.1.3 Publish a Request   The user can choose to publish in the normal way or in a personal way (shown in Figure 4.7).

In personal mode, there is a list showing the user 10 destinations popular in the relevant city. This is achieved by **ListView** which is a view group displaying a list of scrollable items. The user can scroll the view and select the destinations he or she wants.

In the "adapter" directory, there are Java files responsible for creating "adapters," as shown in Figure 4.8.

The adapter Java files act as adapting data resources and **ListView**, transforming the data into a format that **ListView** can display (by attaching the adapter to the **ListView**).

The efficiency of adapters is enhanced by making use of **ListView**'s cache and implementing the ViewHolder class to display the cached contents, which avoids calling methods to find widgets in many times. The main steps are as follows.

- Create bean object to encapsulate data.

- Initialize list of data in constructors.

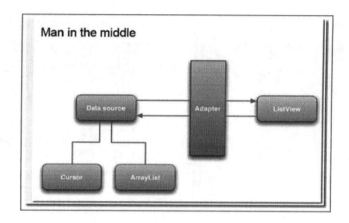

FIGURE 4.8   Adapters.

- Create ViewHolder class to create mapping relationship of the layout.

- Judge convertView, if empty then create a new one, and set a tag, if not empty then get ViewHolder via tag.

- Set data for ViewHolder's widgets.

Then a user needs to enter the destinations or starting point. **InputtipsActivity** is added to help the user find the destination more efficiently and conveniently.

This activity is called when the user need to find the places at which to start and end. It helps the user to find the place more quickly and more efficiently. When something is entered, the **onTextChanged()** function will be called; the application will then query the online server for results. After getting the results, it will handle the results and display on the page, adding the city and the district of the destination. For example, if the user chooses to go to a "fancy" place in Haidian District, Beijing, he or she only needs to enter "f" and look for "fancy" in the list; when the user chooses it, the app will identify the district and city automatically. This will be then stored in the database and listed in the available rides.

The data structure **HashMap** is used to store information on destinations from the database. The **Tip** object is offered by AMAP API, which can be used to get the detailed information on a specific address including the district, the city, its latitude/longitude, etc. When the user confirms the chosen place, it will send it to the former activities and this will be used in the following activities as important data.

After publishing the request, the data will be stored in the **RequestVo** object, which will be sent to the server and stored in the database. The next time the user looks at this page, he or she is able to see all published requests in a scrollable list.

To achieve this, the **AsyncTask** class is used. This puts the preparation work in the main thread inside the implementation of the *onPreExecute()* method, and the *doInBackground()* method will be implemented in the work thread to deal with those heavy tasks. Once the task is finished, the *onPostExecute()* method is called to return to the main thread. Thus we

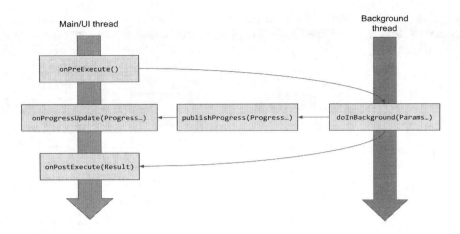

FIGURE 4.9   AsyncTask.

can make interaction between back-end thread and UI thread/main thread. In this case, the tasks communicating to the server can be seen as a heavy task, which is time-consuming; when this operation is finished, we can change or generate new layouts to identity that this task is over. Otherwise, it will cause unexpected errors if we execute some other tasks before the connection work is done (shown in Figure 4.9).

In detail, for the implementation of **AsyncTask** there are three important parameters: *Params* is a required parameter for the background thread. In order to make the background thread work, some necessary parameters need to be offered; for example, for the downloading of a picture, the parameter can be the picture's download address. *Progress* is the progress of the background-thread processing job such as how much the downloading task is completed—20% or 80%, and so on. This number is provided by Progress. *Result* is the result after running the background thread, that is, the information that needs to be submitted to the UI thread—the picture in this example.

4.3.2.1.4 See Available Requests/Take a Ride   To get the request information, **AsyncTask** is also used similarly to adding the data into the database, but in a different direction—from database to Android application.

Figure 4.10 shows screenshots of looking at available requests and taking a ride. When the user clicks the "see available" button on the home page, he or she can see the available requests from people in other roles. For example, if you are a driver, you could see only requests sent by riders. This is achieved by the "type" global variable, which was explained in an earlier subsection.

The user can also see his or her own requests with corresponding status on the Publish page; if the user clicks on a request, the details of the request will display. Then the user can choose to cancel the request or return to the previous page. To implement this, **ActivityDetail** lets the user see all the current requests from other riders and drivers concerning the available rides by using **AsyncTask**.

The user can simply choose an available ride set by the driver and not publish a new one. If the user plans to accept the ride, he or she just click the "accept" button and this will

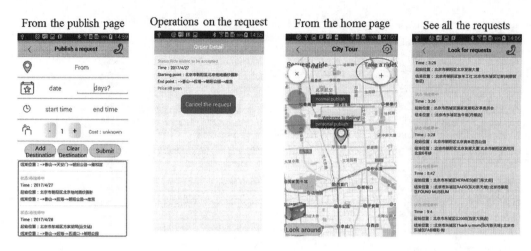

FIGURE 4.10   See available requests/take a ride.

trigger the "status" variable of the request to change. The update of the request status will be seen by users in both roles.

4.3.2.1.5 Start a Trip   If the driver accepts a ride, there is a major divergence. The driver can then click the button "start the ride" on the home page and the app will:

1. Display the price on the right corner

2. Navigate the driver to the destinations (Figure 4.11).

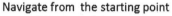

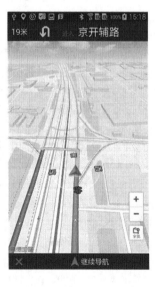

FIGURE 4.11   Start a trip.

This is implemented by **AMAP Navigation SDK** and **MSC SDK**, one for the navigation function and the other for the voice guide function. Using them is the same as described previously: get the key from the official website and then add the necessary files into the development environment. First, **AMapNaviView** is defined in the layout file using **XML**, which displays the navigation view for the user. Second, to ensure the correct display of this navigation view, the implementation life circle is in **Activity_navi**, and in each step of the life circle, different methods will be implemented to ensure the correctness of the logic. After that, the driving route calculation is implemented. There are many strategies for calculating the route; that of avoiding congestion is selected because according to the questionnaire the traveling time is always valued by tourists. To achieve this, there are five parameters in calculating the route in the **strategyConvert()** method:

1. Avoid congestion.

2. Avoid highway.

3. Avoid cost.

4. Highway.

5. Multiple routes.

Setting *false* means not considering this specific strategy. For example, if the developer sets the second parameter to *false*, it means that he or she does not need to calculate a route avoiding highways. It is very easy to change the strategy subsequently.

Finally, after successfully calculating the route, a callback function is called and starts the navigation.

The parameter **NaviType.EMULATOR** is to identify the type of the navigation; it is chosen to use emulator mode because this is convenient to show the functioning of the app. Otherwise, if we remain in a location, the navigation will not continue because we could not move as far or as fast as a car.

4.3.2.1.6 Price and Route Calculation/Tour Plan View   After the user selects the destination and fully fills in the required information, the price and the route will be shown on the page, as illustrated in Figure 4.12.

FIGURE 4.12   Price and route calculation/tour plan view.

The price calculation is similar in the two modes; however, in normal mode the route is fixed—from the starting point to the destination. In the personal mode, the route algorithm will be used to calculate a more efficient route for the user, which is explained in detail later.

DestinationActivity and **PlanActivity** are responsible for implementing this function, one for getting the necessary information and calculating the price and route, and the other for displaying the route and price for the user.

In calculating the price, the time, date, number of people, and places are used. The related data is obtained from user input. Those fields will be complete as there is a verification function to ensure it; otherwise, the user cannot move to the next page. We set the starting price to be 3 yuan and 2.3 yuan per kilometer, which is an average price for taxis.

The price calculation also considers the number of people, since when the weight of the vehicle increases, the fuel cost will be higher. The date and the time will be other factors to include in the calculation of the price. For example, at 8.am and 6.pm (peak hours), an extra 5.5 yuan is added, which is just an estimated increase; on Christmas, Valentine's Day, May Day, and the like, the price will also increase, by different amounts.

In personal mode, the route calculation will be an important task. After the user selects some destinations, an **ArrayList** will store the selected destinations (identified by their IDs), and several **HashMaps** will store the information of the destinations. The information items on destinations is recorded in the SQLite database, such as their priority, longitude, latitude, name, etc.

In the first place, the priority and the location of the selected destinations will be obtained and stored in the corresponding **HashMaps**. This is to calculate the weight of each destination in order to make a route plan. **AMAP API** is used to obtain the distance of each path. With the priority number set in the database, the weight is obtained.

Then, with the use of the Ant Colony algorithm and the Dijkstra algorithm, the route is obtained and the order will be stored in an **ArrayList** called "ordernum".

Next, after obtaining the order of destinations, the visiting time and driving time of each destination are calculated. The driving time is an estimate ride time based on the distance and the average car speed. It is true that the time can be established more accurately by implementing a specific algorithm, but it is not the main focus of this project, so this app uses the estimated time. To calculate the visiting time, the priority of the destination is used. This is based on the percentage of the overall priority. Note that in order to calculate the visiting time of each destination, an upper limit and floor limit are set, which here are 3 hours and half an hour, respectively. The plan list can then be generated. In summary: the total visiting time is calculated from the input of users. The user will give the start time and end time of a day, as well as the starting date and how many days he or she would like to stay. The plan list will then be calculated based on this, including the visiting time of a specific place, the order of the plan and the time of arrival of each place.

4.3.2.1.7 Explore the City   Other functions are added for the user to explore the city including the food, weather, places nearby, etc. (as shown in Figure 4.13).

Activities include: PoiAroundSearchActivity, ActivityTravel, ActivityFood, and WeatherActivity. These are activities responsible for extra tourism activities. For example,

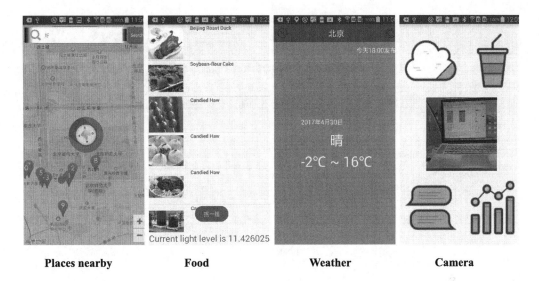

| Places nearby | Food | Weather | Camera |

FIGURE 4.13   Explore the city.

in **PoiAroundSearchActivity**, the user can select a specific place and the application will display the top 10 related places nearby based on the information the user gives. This is implemented using **AMAP API**, firstly initializing the map, then calling query functions and getting the results from the map server, and finally displaying the results on the map.

Also (e.g. in ActivityTravel, ActivityFood, WeatherActivity), a real-time weather searching function and multiple entertainment functions are added to allow users to enjoy the city tour more deeply. For example, added sensor functions include a light sensor, an acceleration sensor, and an orientation sensor. This is implemented using SensorManager, which is the manager of all sensors in an Android system. Application can get any sensor by calling **getDefaultSensor()** methods.

With the "camera" function, users would be able to take a picture and crop it to the size of their preferences. This might be extended further into two parts:

1. Enhance the picture to allow them to share their trip.

2. Use an added artificial intelligence tool to identify the place where they took the picture.

These features are not the main focus of this chapter, and are not yet fully developed, and so will not be considered in detail.

4.3.2.2 Back-End Implementation

The back-end server is implemented by MyEclipse + Tomcat + servlet + MySQL. The data format uses JSON, the protocol uses HTTP. The network environment is that the mobile phone and the server connect to the same wireless router.

In the implementation of the back-end server, an approach named DAO (data access object) design pattern is used, which has two main features:

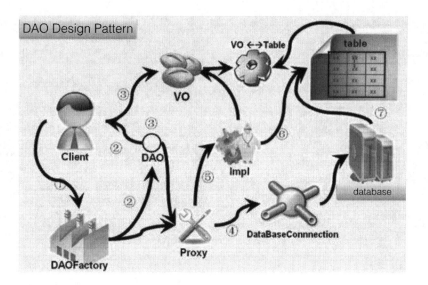

FIGURE 4.14 DAO design pattern.

1. *Layering*: the underlying data logic and the high-level business logic layer are layered in order to achieve decoupling. Figure 4.14 shows an overview of DAO design pattern using in this project.

2. *Data encapsulation*: that is, the data transfer object in DAO components.

DAO design pattern is very important for the implementation of the back-end system, which is actually a combination of two modes, the Data Accessor mode and the Active Domain Object mode. Data Accessor mode is to achieve the separation of data access and business logic; Active Domain Object mode is to achieve the encapsulation of the business data object. The DAO design pattern is a design pattern in Java EE.

DAO has several important components:

- *DAO interface*: DAO is a typical interface-oriented programming. The relationship between classes is through the interface associated with it, rather than the specific implementation. The interface defines the operations in the database.

- DAO interface implementation class completes the above operations defined in the DAO interface. The basic operations in the database, such as connecting the database, can be encapsulated into a specific class.

- *Data Transfer Object Class (value object, VO)*: The data object for mapping into the rows in the database.

4.3.2.2.1 Back-End Server   MyEclipse is an extension of EclipseIDE, which can be used in the database and J2EE development, including publishing capabilities and application server integration to improve the development efficiency. In short, MyEclipse is an Eclipse plug-in

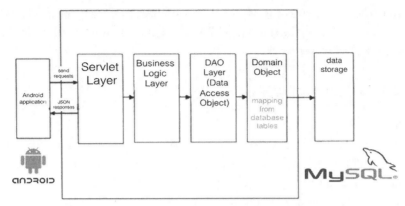

Backend Server Design and Implementation

FIGURE 4.15   Back-end system implementation.

that is also a powerful JavaEE integrated development environment, offering support for code preparation, configuration, testing, and debugging. Tomcat is a JSP/Servlet container, the use of which is suitable for small and medium-sized applications; it is therefore it is suitable for this project.

The architecture of the back-end server is shown in Figure 4.15. The server uses Java EE hierarchical structure, dividing it into view layer, controller layer, business logic layer, and DAO layer. The hierarchical system puts business rules and data access into the middle layer processing. That is, the client does not interact with the database in a direct way but through the controller as well as the middle layer to set up a connection. The middle layer then makes an interaction with the database to access the data.

The servlet class in the server is responsible to response to user requests, getting request parameters and calling business logic layer to handle the requests. Based on the handling results, it will generate the output back to the application and send the encapsulated data in JSON format. In short, the servlet layer does not enter anything or do any processing but is only responsible for receiving, in this project, HTTP requests from the client. It decides which model to call to handle requests, and then determines which data to return.

The server is identified by URL. We choose to use the common approach of REST (representational state transfer) web services. The application makes HTTP requests to the server, sending and receiving data in JSON format. The data will be stored in an entity object and there exits one class in both Android Studio and MyEclipse, which helps to parse JSON. Once the server gets the data, it will parse it and then makes queries to the database, responding to the application after getting the results. This is achieved by JDBC, and we will explain it in the next section.

There are two very common formats for transferring data on the network, XML and JSON. The main reason we choose JSON is that it is smaller and more readable on the network. There are many ways to parse JSON data; we choose to make use of Google's open source library GSON, which can automatically map a JSON-format string into an object, so this project does not need to manually write Java code to parse the data.

In addition, there are two kinds of tool classes in the server side, one for data storage and the other for encapsulating the data into JSON format. Finally, the web.xml file should be deployed to make the server work.

4.4 RESULTS AND DISCUSSION

4.4.1 Results

After the design and implementation part, a working app with the baseline functions—for example, a map of points of interest, selection of points of interest, optimal routing calculation, cost calculation with the help of AMAP SDK/API—was successfully made. Although the specification did not require certain functions, more features were added for tourism purposes, including place search, recommendation for food and shopping places, real-time weather searching, and so on.

The whole system is implemented as **APP-Server-Database**. The Android app is implemented with user-friendly layouts from the beginning, and it sends HTTP client requests to and receives responses from the server. The server connects to a MySQL database, which will return to users whatever they request and display the results.

Surveys from the Chinese government suggest that there will be an increasing number of people choosing individualized travel, but the planning of trips has been a very troublesome and time-consuming process for the individual. It was in this context that that the travel assistance tool was born. Most tour apps currently on the market rely on the most basic way to introduce the user to the trip—manually. The trip assistant tool described here is an intelligent product rather than a simple production tool.

The user only needs to determine the travel destinations and the app will be able to intelligently help users to calculate the route at a reasonable price and time. The user can refer to the plan and adjust it to a full tour plan based on their own preferences. This can save a lot of time for tourists and it offers drivers or tour operators multiple business opportunities.

4.4.2 Discussion

4.4.2.1 Testing and Modification

The application was tested constantly during the implementation phase. The purpose was to find errors and bugs and to solve them so as to make the app more reliable. For Android applications, it may run on various devices so it is important to test on different devices. The application was first tested in devices with lower configurations and then on devices with higher configurations (e.g., screen resolution, to ensure that is will work adequately on different Android mobile phones).

The test then focuses on the business logic of the application; this project breaks down the test into three parts:

- Ensuring stability of the code basis.

- Functional tests to ensure that the app will work successfully.

- Cross-functional tests.

TABLE 4.3   Test Cases of the Application

Test Mode	Case No.	Input	Output
User login/register	1	Wrong password	An alert message will display
	2	Wrong format of password	An alert message will display
	3	No role selected	An alert message will display
Publish a request	1	Leave some fields empty	An alert message will display indicating the field that is empty
	2	Wrong format of the number of people	An alert message will display indicating the format is wrong
	3	Cancel the destination	Cancel successfully
Accept a ride	1	Click button "accept"	The status of the ride will change
	2	Return back	Back to the home page
Search places nearby	1	Input nothing	An alert message will display indicating that the field is empty
	2	Input wrong place	An alert message will display indicating that there exists no place of this name
Modify the status of the ride	1	Cancel the ride	The status of the ride will change
	2	Accept the ride	Store successfully and change the status

The functional tests are based on black-box testing to verify functionality according to the specification. Table 4.3 shows the test cases of the application.

One main error occurs during testing phases, which is the synchronization problem. To send the data to and get the data from the database, the thread needs to perform some time-consuming tasks and will block the UI thread that displays the page for the users. Therefore, we add an inner class which extends **AsyncTask** to solve this problem.

4.4.2.2 User Reviews

Persons outside the project were invited to test the application, and their experience helped to identify weakness that had not been noticed: when the user clicked the back button of the mobile phone, the plan list displayed repeated destinations. This was caused by not implementing the **onBackPressed**() method. The code was modified to handle the logic when the user clicks the back button, and the app now performs well.

Other problems that were identified include the lack validations of user input. For example, when the user entered a wrong format for a mobile phone number, the app did not display any error message. Subsequently, numerous format validations were added to the app to maintain its reliability. Also, functions including "remember me" were added, which store the password of the user so that next time the user logs in, he or she does not need to enter it again.

4.5 CONCLUSION AND FURTHER WORK

4.5.1 Conclusion

A working app for individualized trips has been achieved and the app is able to support a basic framework for such use. All the necessary functions were considered and were implemented to create the app's user interaction that draws and receives events. The route calculation algorithm is based on the Dijkstra algorithm with some alterations. The price calculation algorithm considers four factors including distance, traffic, special time, and special days. With further development the project will optimize these algorithms and the distribution of the visiting time for each destination. This project has followed UI design based on the responses from interviews and questionnaires to complete the links among different pages and during development continually improved the UI design to make the app more attractive and user-friendly. In addition, the app has some extra functions more appropriate to tourism, such as point of interest search and route navigation.

4.5.2 Challenges and Solutions

We started the development of this app after reviewing the basic features of Android programming. However, throughout the process we encountered numerous difficulties concerning Java multithreading. At first we put too many tasks into the main thread and as a result not only the main function was affected. And as a result of that problem, there were run-time errors that caused the app to stop.

In seeking to solve the problem, we consulted tutorials online and the official documentation of Java to acquire knowledge for programming in Android. The obvious place to begin was the official site for Android developers offered by Google. The Stack Overflow website was also searched. Various problems were resolved after examining similar questions and their answers. For example, the multithreading part of the app operates frequently in this app to query the back-end system and access the data. A solution called **AsyncTask** is offered by Stack Overflow and some other online tutorials that enables correct and easy use of the UI thread. It helped a lot to solve the problems, allowing this app to perform background operations and publish results on the UI thread without having to manipulate threads or handlers.

4.5.3 Future Work

In the future, as the data accumulate at a larger scale, this application project may inevitably turn to machine learning—not only to implement common machine learning algorithms but perhaps to design a new one based on the specific requirements of the app. The accumulated data would be very useful in optimizing the efficiency and accuracy of the app. For example, Amazon cannot neatly encode the tastes of all its customers in a computer program, and Facebook does not have a program that will choose the best updates to show to each of its users. This project will turn learning algorithms loose on the mountains of data they have accumulated and let them divine what customers want.

Certainly some features of this application can potentially be improved using machine learning, for example to detect users' private preferences and interests, which is an essential part of individualized tours. Also, it will be required to keep machine learning algorithms running and continuously adjust many aspects of the application. Application of machine learning techniques to the establishment of individual tour preference will improve the ability to make personalized recommendations.

The price calculation algorithm in particular could usefully be based on a machine learning linear regression algorithm, which is a linear model that assumes a linear relationship between the input variables and the single output variable. It would be supervised learning for, in this project, ride pricing. Numerous techniques exist because the model has been extensively studied. According to the Machine Learning Mastery website [22], "learning a linear regression model means estimating the values of the coefficients used in the representation with the data that we have available."

This project has already found a way to implement the machine learning method by using Weka, Java library, which provides machine learning algorithms for tasks this project needs to handle. The algorithms can either be applied directly to a dataset or called from the Java code.

A main advantage of machine learning algorithms is that the accuracy increases as the amount of data to learn from increases, which suits the purposes of this project. Also, both methods will double the resource utilization, which is undoubtedly a benefit to overall optimization. For example, the priority rating of each location is set in the database instead of being dynamically generated or updated over time. However, this can be improved by using online ratings data from well-known and well-used travel websites or updated according to machine learning algorithms. As the data is growing, the planning can be generated by supervised or unsupervised learning based on machine learning algorithms, and even the algorithm itself can be chosen using other algorithms. Data is like a gold-mine that will help us to explore the world further.

BIBLIOGRAPHY

1. Griffiths, D., and Griffiths, D. 2015. *Head First Android Development.* O'Reilly Media, Inc., Newton, MA, USA.
2. Hamilton, G., Cattell, R., and Fisher, M. 1997. *JDBC Database Access with Java* (Vol. 7). Addison Wesley, Boston, MA, USA.
3. Chi, Z. 2016. *China's Tourism Statistics Bulletin 2015.* Retrieved January 24, 2017, from China National Tourism Administration. http://www.cnta.gov.cn/zwgk/lysj/201610/t20161018_786774.shtml
4. Android Official Guide. 2016. *Official Guide for Android Development.* Retrieved October 21, 2016, from Google Android Developer: developer.android.com
5. Ouyang, H., Xie, Z., Huang, S., and Ding, Y. 2009. A data persistence layer model based on DAO design pattern and hibernate framework. *Microcomputer Applications*, 3, 36–40. doi: 10.3969/j.issn.2095-347X.2009.03.008
6. Kwak, D., Kim, D., Liu, R., Nath, B., and Iftode, L. 2015. DoppelDriver: Counterfactual actual travel times for alternative routes. In *2015 IEEE International Conference on Pervasive Computing and Communications (PerCom)*, (pp. 178–185). IEEE, Piscataway, NJ, USA.
7. Romeing, H. 2015. *UberTOUR: Sightseeing The New Way.* Retrieved October 12, 2016, from Romeing: http://www.romeing.it/rome-ubertour-sightseeing.
8. Li, Z., Zhu, Y., and Li, M. 2009. Practical location-based routing in vehicular ad hoc networks. In *IEEE 6th International Conference on Mobile Adhoc and Sensor Systems, 2009. MASS'09*, (pp. 900–905). IEEE, Piscataway, NJ, USA.
9. Smirnov, A., Shilov, N., and Gusikhin, O. 2016. "Connected Car"-based customised on-demand tours: The concept and underlying technologies. In *International Conference on Next Generation Wired/Wireless Networking*, (pp. 131–140). Springer International Publishing, Cham, Switzerland.
10. AMAP Official Guide. 2015. *Using AMAP API for Android Development.* Retrieved October 27, 2016, from AMAP API: http://lbs.amap.com/api/android-location-sdk/locationsummary

11. Zapata, B. C. 2013. *Android Studio Application Development*. Packt Publishing Ltd., Birmingham, UK.

12. Chang, W. C., Tai, H. T., Hsieh, D. L., Yeh, F. H., and Chang, S. H. 2013. Design and implementation of the travelling time- and energy-efficient android GPS navigation app with the VANET-based A* Route Planning Algorithm. In *2013 International Symposium on Biometrics and Security Technologies*, Chengdu, China (pp. 85–92).

13. Liu, S., Leng, H., and Han, L. 2017. Pheromone model selection in ant colony optimization for the travelling salesman problem. *Chinese Journal of Electronics*, 26(2), 223–229.

14. Zheng, J. J. 2016. *Navigation from one place to multiple destinations for the best route*. Retrieved March 25, 2017, from Wan Fang Data Library: http://g.wanfangdata.com.cn.

15. Rani, C. R., Kumar, A. P., Adarsh, D., Mohan, K. K., and Kiran, K. V. 2012. Location based services in android. *International Journal of Advances in Engineering & Technology*, 3(1), 209–220.

16. Bender, M., and Westphal, S. 2015. An optimal randomized online algorithm for the k-Canadian Traveller Problem on node-disjoint paths. *Journal of Combinatorial Optimization*, 30(1), 87–96.

17. Yurchak, W., and Dudas, V. 2017. Investigation into applicability and efficiency of SQLite for implementation of Dijkstra's algorithm. In *2017 14th International Conference The Experience of Designing and Application of CAD Systems in Microelectronics (CADSM)*, Lviv - Polyana, Ukraine, (pp. 282–284).

18. Welling, L., and Thomson, L. 2003. *PHP and MySQL Web Development*. Sams Publishing, Carmel, IN, USA.

19. Monseul, M. 2016. *An Implementation of Ant Colony Optimization for TSP problem in Java*. Retrieved March 12, 2017: http://blog.csdn.net/wuchuanpeng/article/details/51583829

20. Saini, L. M., Aggarwal, S. K., and Kumar, A. 2010 January. Parameter optimization using genetic algorithm for support vector machine-based price-forecasting model in national electricity market. *Generation Transmission & Distribution IET*, 4(1), 36–49.

21. Guo, L. 2014. *The First Line of Code: Android*. The People's Posts and Telecommunications Press, Beijing, China. In Chinese.

22. Brownlee, J. 2013. *A tour of machine learning algorithms. Machine Learning Mastery*. Retrieved January 11, 2016, from Machine Learning Mastery: http://machinelearningmastery.com.

23. *Daily FT*. 2016. London sets record with over 31.5 million visitors in 2015. http://www.ft.lk/article/543880/London-sets-record-with-over-31.5-million-visitors-in-2015

Teaching Hygiene and Responsible Antibiotic Use through a Mobile Game for Children

Andreea Molnar and Patty Kostkova

CONTENTS

5.1 INTRODUCTION

Serious games and gamification have been identified as one of the main challenges of current pursuit in digital health [17]. Educational games that teach through the mechanics of the game have been evaluated as an effective educational intervention [40]. However, designing games aimed at children has its own challenges and issues [30].

Raising the awareness of and better public and professional education about the global challenge of antimicrobial resistance is one of the goals in attempts to reverse this dangerous global trend of growth in resistance [24]. Public attitudes towards prescribing are seen as an important factor in antibiotics overprescribing [7] that could be changed by targeted internet websites such as *Bugs and Drugs* [22] and *Do Bugs Need Drugs?* (http://www. dobugsneeddrugs.org/).

However, targeting children is of particular importance. Teaching hand and respiratory hygiene to reduce illness (which that could potentially require antibiotics) among children has proved to have strong impact on children's health and school absences [11,39] as they will educate the future generation of antibiotics users.

This chapter will present a mobile game that it is aimed at 9- to 12-year-olds and teaches about healthcare issues based on the core concepts from the European curriculum [21]. It will expand on the implementation challenges of developing and porting a mobile game compared with a previous desktop-based solution [31]. It will discuss the learning objectives covered in the game and how they have been integrated into the game mechanics to enable assessment through the method of "seamless evaluation". The chapter will also present the results of a study assessing the game's usability.

Technically we will present the issues encountered while developing a mobile game as opposed to a desktop-based game and how they have been addressed. Most of these issues are not unique to our application and our approach could be useful for other projects. We also consider how the learning objectives could be integrated into the game mechanics, which could be relevant for other serious games projects.

5.2 RELATED WORK

Recently, serious games have become well established as a new method of education. The popularity of games and the "motivational power of games", as well as the opportunity to harness the learning that happens in games, has led to an interest in the development of games for education [14]. Recently, serious games have become well established as a new method of education.

This new trend has not been without its challenges. The results of "edutainment" (or educational entertainment) have been disappointing. Kirriemuir and McFarlane [14] suggested that educational games continue to fail when competing with commercial games due to the former's simplistic, repetitive, and poor design of the mechanics. However, the positive use of games to aid learning has been shown in different studies [37,41]. Games have been successfully used to teach about microbe transmission and the importance of hand washing [27], to train grapho-motor skills [36], and to improve knowledge of medieval history [23].

Serious games and storytelling provide the users with an engaging and motivating experience [10]. The game stories situate learning in a context [13], and interactivity allows the player to be an active actor in the story construction and may lead to unintentional learning [10].

There is indeed potential for applying interactive storytelling to learning [15]; however, evaluation of the educational impact of the game against a set of learning objectives (LOs) remains a challenge.

The effectiveness of serious games in terms of learning outcomes is still understudied, mainly due to the complexity involved in assessing intangible measures [3]. To assess the educational impact of games, novel approaches are required that fully take advantage of the interactive environment and game mechanics while not decreasing user immersion and engagement with the game. In-game interaction data have been used to improve assessment of a game in the eAdventure platform [38].

To make full use of the in-game data, an integration of the LO evaluation into the game mechanics—the so called "seamless assessment" or "seamless evaluation" framework [28]—has been developed and is in use. The games showed the ability to change based on the improvement in student knowledge. Moreover, including the feedback of the assessment in the score or the interactive digital storytelling reward system can act as an extrinsic motivation for the players and increase enjoyment of the game. While the SUS (system usability scale) score framework has been established in the domain for assessing usability [2], an assessment of knowledge change and improvement throughout by seamlessly collecting answers to questions in the game has been developed by Kostkova [16] and Molnar and Kostkova [28]. The study of user engagement with digital health technologies identified approaches to increasing user engagement with serious games while assessing knowledge change and retention [18].

5.3 EDUGAMES4ALL MICROBEQUEST!

5.3.1 Edugames4all

Edugames4all is a collection of digital games aiming to improve the learning of children (9–16 years old) about important issues such as hygiene, microbe transmission, and responsible antibiotic use. The games' educational content is based on the learning objectives taught across the European curriculum [21]. The collection consists of two different types of games: platform games [9,29] aimed at 9- to 12-year-olds; and interactive digital storytelling-based games aimed at young adults 13–16 years old [32,33].

Interactive digital storytelling-based games are designed following the STAR framework [27]. The games teach about the importance of handwashing and responsible antibiotic use among young adults. They have been evaluated in terms of both their educational potential [16,28] and their usability and player enjoyment [20,32,33].

The platform games are designed for 9- to 12-year-olds. They were designed to meet children's preferences for this age group. The games teach about microbes, hand, and food hygiene, and responsible antibiotic use [9,29]. The games teach the LOs through a combination of text and game mechanics [29]. As for previous games, their usability and educational potential have been assessed [9,32,34].

Although most of these games were originally designed to be played online from a desktop or laptop, one of the games was modified to be played on mobile devices [31]. The mobile version of the game, edugames4all MicrobeQuest!, is presented below.

5.3.2 Software Development Life Cycle

Here we describe edugames4all MicrobeQuest! by broadly dividing the work into the first three steps of the software development life cycle: inception, design, and development. Stabilization is presented in the following section.

5.3.2.1 Inception

The addition of a mobile version of the game was motivated by children's increased access to mobile devices. Edugames4all MicrobeQuest! was developed as a mobile version of one of the edugame4all platform games. The game has several missions, with each mission focusing on one aspect of three themes: microbe transmission, food and hand hygiene, and responsible antibiotic use. Each mission focuses on a certain environment: hands, food, or body. To see the microorganisms, in some missions, the player shrinks to the size of a microorganism and explores the environment. The learning objectives covered in the game along with the environment in which the learning objective is presented are shown in Table 5.1.

As with previous games, the learning objectives covered in the game are based on learning objectives taught across several European countries. The learning objectives aim to familiarize the children with microorganisms, their transmission, their role in hygiene, and the importance of using the antibiotics only when and as prescribed by the healthcare professional.

5.3.2.2 Design

The game design is based on the desktop version of the game and follows relatively closely the desktop version of the platform games [31]. Prior to development of the desktop version of the game, design focus groups were held with primary school children to determine the type of game most suitable for this age group [8]. From the results of the focus groups, it

TABLE 5.1 Learning Objectives Covered in the Games and the Environment They Relate to

Learning Objective	Environment
Bacteria, viruses, and fungi are three different types of microbes	Hands, Food and Body
Bacteria, viruses, and fungi can be found in different environments	Hands, Food and Body
Bacteria, viruses, and fungi come in different shapes and sizes	Hands, Food and Body
"Bad" microbes can make us ill	Hands, Food and Body
Not all microbes are harmful	Hands, Food and Body
Washing can eliminate harmful microbes	Hands
Useful microbes can make food like yogurt (and bread)	Food
Our bodies have defences to fight off disease	Body
Antibiotics can be used to fight off bacterial infections	Body
If antibiotics are taken, it's important to finish the course	Body
Vaccines can be used to obtain immunity to viral diseases	Body
Vaccines can be used to obtain immunity to viral diseases where the body's natural defences alone are not enough	Body
Bacteria are becoming resistant to many antibiotics due to antibiotic misuse (not finishing the course)	Body

was decided that platform games were the type that children enjoyed [8]. Furthermore, other focus groups and observational studies were performed to validate the characters in the game and determine what it is the best way to present the children the characters in the games [8].

At the beginning of the game, the player will choose an avatar. Afterwards, in most of the game missions, the player will shrink to the size of a microorganism and explore different parts of the body or the kitchen. Through the interaction with the microorganism, the player is taught several learning objectives that focus on microbial transmission, hand and food hygiene, and antibiotic resistance. The learning objectives are taught through a combination of text and game mechanics [29,34]. Our previous work [29] has shown that it not always clear whether teaching with text or with game mechanics will produce better results, and in this case we opted to use both approaches, one method reinforcing the other. For example, teaching about the "good microbes" and some of their applications through text is an example of the message provided by the game to the players: "*Good microbes can turn milk into yoghurt. That's how yoghurt gets made. Amazing isn't it?*". Through game mechanics, the player must push a *Lactobacillus* bacterium into a glass of milk. When the bacterium touches the milk, the milk becomes yogurt.

MicrobeQuest! allows measurement of the player's knowledge before and after they tackle the learning objectives through the game. Teachers and the players can thus determine how much their knowledge has improved because of playing the game. The assessment implementation is done through a quiz such as "*How to be a millionaire*". The learning objectives that are to be covered in the next levels are assessed at the beginning of the game and no feedback is provided to the player whether their answers were accurate. After each level, the learning objectives covered in the previous level are assessed through the same quiz, but now the players are provided with feedback on their answers. Previous research has shown that this kind of assessment does not affect how the players perceive the game and most of the players preferred it as opposed to being assessed through a questionnaire [19].

5.3.2.3 Development

To allow the game to be played across different platforms we used a hybrid approach in developing the mobile game. During the process, we encountered several challenges specific to the mobile version of the game [31] when trying to transform the desktop version of the game into a mobile one.

First, mobile devices have a multitude of platforms with different requirements. A hybrid approach to the development means making different trade-offs in terms of rendering and graphics. For testing, we decided to focus on optimizing the application for one mobile operating system—Android—as this is the most commonly used operating systems in terms of internet usage [5] and can reach as many players as possible.

Second, the appropriateness of a development platform in the rapidly changing mobile market remains a challenge. The desktop versions of the games were developed in Flash due to the availability of the plugin on computer internet browsers in most schools [9] at the time when the games were developed. However, mobile device support for Flash is declining. As a result, it was not an adequate technology to use in developing the games and especially games for mobile devices. As both the availability of HTML5 and JavaScript and their capability is increasing along with support across different platforms we decided to implement the games in HTML5 and JavaScript.

Third, for mobile applications small size is important due to the memory constraints but also because of limited bandwidth available for mobile applications. When the desktop application was built, we considered that the internet connection might fluctuate and therefore tried to keep the size of the games small. Nevertheless, most of the assets still had to be modified for the mobile device to keep the mobile application size light.

Fourth, the internet connectivity on mobile devices cannot be assured, and sometimes users might have concerns regarding the data usage and the costs associated with it [26]. Edugames4all MicrobeQuest! uses internet connection only if the player explicitly agrees to it when using the app (see Figure 5.1). The internet connection is used to deliver the players' scores and retrieve the other players' scores, for the player to be able to compare his or her performance with that of other players. The players' score is transmitted to the server only if the players agree to it. To allow this feature, when there is no available internet connectivity, the scores are stored locally until the next time the player uses the game and has an internet connection. This approach also allows users with no or limited data plans over 3G and 4G protocols (common for children's mobile use) to enjoy the gaming experience without worrying about the availability of free Wi-Fi hotspots or other means of affordable connectivity.

Fifth, mobile devices screen sizes vary, and different screen sizes would allow for a different amount of information to be displayed. To avoid affecting the player experience we used a "zoomed-in" approach for games played on small devices and "zoomed-out" for games played on larger screens. Figures 5.2 and 5.3 present two examples where this approach was implemented.

Table 5.2 presents a summary of the development challenges we encountered and a brief description of how they have been addressed.

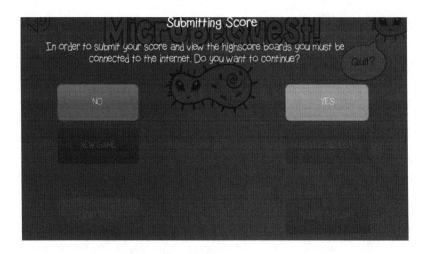

FIGURE 5.1 Explicit internet connection.

FIGURE 5.2 Player exploring the kitchen: game screenshot for 800 × 480 px device.

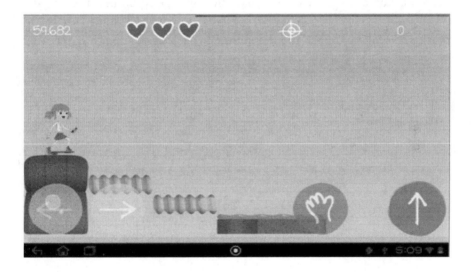

FIGURE 5.3 Player exploring the kitchen: game screenshot for 1280 × 720 px device.

TABLE 5.2 Summary of Development Challenges

Challenge	Adopted Solution
Different mobile platforms	Hybrid application development
Declining usage of Flash	Use of HTML5 and JavaScript for development
Memory constraints	All art was adapted to the mobile device specification
Internet connectivity, especially among children	The game saves data locally and sends it to the server only when the user explicitly agrees for it to be sent
Different screen sizes	A "zoomed-in" approach for games played on small devices and "zoomed-out" for games played on larger screens

5.4 STABILIZATION

In the stabilization phase we aimed to assess the usability of the system. We used mixed methods to assess the usability of the game. System Usability Scale (SUS) [4] was used to quantitatively assess the usability of the game, while observations were used to collect qualitative data and further feedback. Observations could be used when observing a phenomenon [25], in our case the participants playing the game. The observations were done by the researcher who conducted the study.

SUS is a 10-question usability questionnaire used to assess and quantify perceived usability. The questionnaire is reliable and has been validated [1]. It has been used in the past to assess usability in different contexts, including games (see, for example, 12,33,35,42). The original questionnaire was modified so that it was adapted to assess a game rather than a system in general as described in reference [33]. The word "system" has been changed to "game" and "use" changed to "play". Moreover, some of the words were modified to be suitable for participants in this age group. The final version of the questionnaire is presented in Table 5.3. The SUS uses a Likert scale. The participants must select from a scale from 1 to 5, where 1 is labelled with *Strongly Disagree* and 5 is labelled as *Strongly Agree*. Based on the answers, a usability score can be computed and the result interpreted based on the Bangor et al. scale [2]. However, the SUS is used to quantify the usability and not to explain why problems in usability occur and therefore was used together with observations to get a richer overview of the usability of the game.

5.4.1 Participants

A total of 29 participants from different socioeconomic backgrounds and ages participated in the study on a voluntary basis. The participants were selected from two UK primary schools and an afterschool club. Their ages were between 8 and 11 years old, with an average age of 9.5 years old. A total of 15 participants were females and 14 were males.

5.4.2 Set-Up

The children were provided with a mobile phone (Samsung Galaxy S4). The game was pre-installed on the mobile phone. They were asked to play the game for 30 minutes. Afterwards

TABLE 5.3   System Usability Scale (SUS) Questionnaire

Question No	Question Statement
1.	I think I would like to play this game frequently
2.	I found the game too complicated when it did not need to be
3.	I though the game was easy to play
4.	I think I would need the support of a teacher or other expert to be able to play this game
5.	I found the various functions fitted well in the game
6.	I thought there was too much inconsistency in this game
7.	I would image that most people would learn to play this game quickly
8.	I found the game very awkward to use
9.	I felt very confident playing the game
10.	I need a lot of help before I could get to play this game

they were asked to fill a questionnaire which include questions asking about: their age and gender and the SUS questionnaire. After the study the children participated in a brief focus group with the researcher present. The focus group focused on the issues observed during the game play session.

5.4.3 System Usability Scale

The SUS score of the game was 71, which according to Bangor et al. [2] is "good" usability. According to the same study, this usability is also considered "acceptable" (on a scale consisting of "not acceptable", "marginal" and "acceptable").

5.5.4 Observations and Focus Group

Not all the children succeeded in advancing to level 2 of the game and very few managed to finish the game. The ones who did not manage to progress had difficulties executing the game mechanics (a game mechanics which did not pose difficulties on the desktop version of the game). The children did not have a problem with understanding the game mechanics, but the execution on the mobile phone posed challenges. Some of them also felt that the mobile game was too slow. Both these problems were found regardless of whether the children had played a similar game in the past or not.

5.5 SUMMARY

This chapter presented edugames4all MicrobeQuest! [33], a game aimed at teaching children about hygiene and responsible antibiotic use. The challenges developing the mobile version of this game are discussed. The chapter also presents the learning objectives covered in the game and how they have been integrated into the game mechanics. A usability study was performed following a mixed method approach [6]. The results of the study showed that the game usability is good, children finding the game overall easy to learn and use, but the game mechanics employed has been found difficult to execute on the mobile phone.

We learned several lessons from this study. First, some of the game mechanics could not be translated directly to the mobile phones from the desktop version. Second, the children's speed of playing the game might be different between the mobile and desktop version of the game. Third, catering for different platforms and screen sizes required adaptation of the game. Fourth, the assets developed for the desktop version of the game needed considerable adaptation for the mobile version. Fifth, the internet connection, which is often guaranteed when the games are played from the desktop browser, could pose challenges on the mobile devices.

In our future work, we want to explore the game's potential to teach children and change behaviour. With the growing usage of mobile phones among children, we are also exploring performing an in-depth study which considers children's differences in interaction between the desktop and mobile version of the game with focus on improving the engagement with the mobile version, and how these affect usability, enjoyment, and learning.

REFERENCES

1. Bangor, A., Kortum, P., and Miller, J. A. 2008. The system usability scale (SUS): An empirical evaluation. *International Journal of Human-Computer Interaction*, 24(6), 574–594.

2. Bangor, A., Kortum, P., and Miller, J. 2009. Determining what individual SUS scores mean: Adding an adjective rating scale. *Journal of Usability Studies*, 4(3), 114–123.

3. Bellotti, F., Kapralos, B., Lee, K., Moreno-Ger, P., and Berta, R. 2013. Assessment in and of serious games: An overview. *Advances in Human-Computer Interaction*, 2013, article no. 1.

4. Brooke, J. 1996. SUS-A quick and dirty usability scale. *Usability Evaluation in Industry*, 189(194), 4–7.

5. Business Wire. 2017. *Android Overtakes Windows for First Time – StatCounter*. Available online: http://www.businesswire.com/news/home/20170403005635/en/Android-Overtakes-Windows-Time-%E2%80%93-StatCounter (Accessed October 27, 2017).

6. Creswell, J. W., and Clark, V. L. P. 2007. Designing and Conducting Mixed Methods Research. Sage, Thousand Oaks, CA, USA.

7. Eng, J. V., Marcus, R., Hadler, J. L., Imhoff, B., Vugia, D. J., Cieslak, P. R., and Hawkins, M. A. 2003. Consumer attitudes and use of antibiotics. *Emerging Infectious Diseases*, 9(9), 1128.

8. Farrell, D., Kostkova, P., Lazareck, L., Weerasinghe, D., Weinberg, J., Lecky, D. M., and Merakou, K. 2011a. Developing e-Bug web games to teach microbiology. *Journal of Antimicrobial Chemotherapy*, 66(suppl_5), v33–v38.

9. Farrell, D., Kostkova, P., Weinberg, J., Lazareck, L., Weerasinghe, D., Lecky, D. M., and McNulty, C. A. 2011b. Computer games to teach hygiene: An evaluation of the e-Bug junior game. *Journal of Antimicrobial Chemotherapy*, 66(suppl_5), v39–v44.

10. Hodhod, R., Cairns, P., and Kudenko, D. 2011. *Innovative Integrated Architecture for Educational Games: Challenges and Merits*. Springer, Berlin, Heidelberg, pp. 1–34. In Pan Z., Cheok A.D., Müller W., Yang X. (eds) Transactions on Edutainment V. Lecture Notes in Computer Science, vol. 6530.

11. House of Lords. 1998. Resistance to antibiotics and other antimicrobial agents. Select Committee Report on Science and Technology, HL paper no. 81–I. London, UK.

12. Hupont, I., Gracia, J., Sanagustin, L., and Gracia, M. A. 2015. May. How do new visual immersive systems influence gaming QoE? A use case of serious gaming with Oculus Rift. In *2015 Seventh International Workshop on Quality of Multimedia Experience (QoMEX)*, pp. 1–6. IEEE., Piscataway, NJ, USA.

13. Kapp, K. M. 2012. *The Gamification of Learning and Instruction: Game-Based Methods and Strategies for Training and Education*. John Wiley & Sons.

14. Kirriemuir, J., and McFarlane, A. 2004. Literature review in games and learning. Available at: https://telearn.archives-ouvertes.fr/hal-00190453/document. Accessed at 19 September 2018. Accessed at 19 September 2018.

15. Klabbers, J. H. 2003. The gaming landscape: A taxonomy for classifying games and simulations. In *DiGRA Conference*, November 4–6, 2003, Utrecht, The Netherlands.

16. Kostkova, P. 2011, November. Seamless evaluation of interactive digital storytelling games: Edugames4All. In *International Conference on Electronic Healthcare*, (pp. 80–84). Springer, Berlin Heidelberg.

17. Kostkova, P. 2015a. Grand challenges in digital health. *Frontiers in Public Health*, 3(134), doi: 10.3389/fpubh.2015.00134.

18. Kostkova, P. 2015b. User engagement with digital health technologies. In O'Brien, H. and Cairns, P. (Eds): *Why Engagement Matters: Cross-Disciplinary Perspectives of User Engagement in Digital Media*, Springer international, Chams, Switzerland.

19. Kostkova, P. and Molnar, A. 2014. Educational games for creating awareness about health issues: The case of educational content evaluation integrated in the game. In *Medicine 2.0 Conference*. JMIR Publications Inc., Toronto, Canada.

20. Lazareck, L. J., Farrell, D., Kostkova, P., Lecky, D. M., McNulty, C. A., and Weerasinghe, D. 2010. Learning by gaming-evaluation of an online game for children. In *2010 Annual International Conference of the IEEE Engineering in Medicine and Biology Society (EMBC)*, pp. 2951–2954. IEEE., Piscataway, NJ, USA.

21. Lecky, D. M., McNulty, C. A., Adriaenssens, N., Koprivová Herotová, T., Holt, J., Touboul, P., and Campos, J. 2011. What are school children in Europe being taught about hygiene and antibiotic use?. *Journal of Antimicrobial Chemotherapy*, 66(suppl_5), v13–v21.
22. Madle, G., Kostkova, P., Mani-Saada, J., and Weinberg, J. R. 2003. Evaluating the changes in knowledge and attitudes of digital library users. In *International Conference on Theory and Practice of Digital Libraries, ECDL 2003*, pp. 29–40. Springer, Berlin, Heidelberg.
23. Markouzis, D., and Fessakis, G. 2016. Rapid prototyping of interactive storytelling and mobile augmented reality applications for learning and entertainment—The case of "k-Knights". *International Journal of Engineering Pedagogy*, 6(2), 30–38.
24. McNulty, C. A., Cookson, B. D., and Lewis, M. A. 2012. Education of healthcare professionals and the public. *Journal of Antimicrobial Chemotherapy*, 67(suppl_1), i11–i18.
25. Merriam, S. B., and Tisdell, E. J. 2015. *Qualitative Research: A Guide to Design and Implementation*. 4th edition. Wiley/Jossey-Bass, New York.
26. Molnar, A. 2014. On better understanding the usage of mobile phones for learning purposes. *Bulletin of the IEEE Technical Committee on Learning Technology*, 16(2/3), 18–20.
27. Molnar, A., Farrell, D., and Kostova, P. 2012. Who poisoned hugh?-the STAR framework: Integrating learning objectives with storytelling. In *International Conference on Interactive Digital Storytelling*, pp. 60–71. Springer, Berlin, Heidelberg.
28. Molnar, A., and Kostkova, P. 2013a. Seamless evaluation integration into IDS educational games. In *International Conference on the Foundations of Digital Games (FDG 2013)*, pp. 322–329. Chania, Crete, Greece, May 14–17, 2013. Society for the Advancement of the Science of Digital Games, Santa Cruz, CA, USA. ISBN: 978-0-9913982-0-1
29. Molnar, A., and Kostkova, P. 2013b. On effective integration of educational content in serious games: Text vs. game mechanics. In *2013 IEEE 13th International Conference on Advanced Learning Technologies (ICALT)*, pp. 299–303. IEEE., Piscataway, NJ, USA.
30. Molnar, A., and Kostkova, P. 2013c. If you build it would they play? Challenges and solutions in adopting health games for children. *Proceedings of ACM SIGCHI Conference on Human Factors in Computing Systems, Let's talk about Failures: Why was the Game for Children not a Success*. ACM, New York, NY, USA.
31. Molnar, A., and Kostkova, P. 2015. Mind the gap: From desktop to app. In *Proceedings of the 5th International Conference on Digital Health 2015*, pp. 15–16. ACM, New York, NY, USA.
32. Molnar, A., and Kostkova, P. 2016a. Ubiquitous bugs and drugs education for children through mobile games. In *Proceedings of the 6th International Conference on Digital Health Conference*, pp. 77–78. ACM, New York, NY, USA.
33. Molnar, A., and Kostkova, P. 2016b. Interactive Digital Storytelling based educational games: Formalise, author, play, educate and enjoy!-The edugames4all project framework. In *Transactions on Edutainment XII*, pp. 1–20. Springer, Berlin Heidelberg.
34. Molnar, A., and Kostkova, P. 2018. Learning about hygiene and antibiotic resistance through mobile games: Evaluation of learning effectiveness. In *DH'18: 2018 International Digital Health Conference*, pp 95–99. Lyon, France, April 23–26. ACM, New York, NY, USA.
35. Muri, R., and Mosimann, U. P. 2016. Usability assessment of natural user interfaces during serious games: Adjustments for dementia intervention. *Journal of Pain Management*, 9(3), 333.
36. Padilla-Zea, N., Gutiérrez, F. L., López-Arcos, J. R., Abad-Arranz, A., and Paderewski, P. 2014. Modeling storytelling to be used in educational video games. *Computers in Human Behavior*, 31, 461–474.
37. Robertson, J. C., and Howells, C. 2008. Computer game design: Opportunities for successful learning. *Computers & Education*, 50(2), 559–578.
38. Serrano, Á., Marchiori, E. J., del Blanco, Á., Torrente, J., and Fernández-Manjón, B. 2012, April. A framework to improve evaluation in educational games. In *2012 IEEE Global Engineering Education Conference (EDUCON)*, pp. 1–8, IEEE., Piscataway, NJ, USA.

39. SMAC - Standing Medical Advisory Committee, Sub-group on antimicrobial resistance. 1998. *The Path of Least Resistance*. Department of Health, London, UK. Available at http://webarchive.nationalarchives.gov.uk/20081106020107/http://www.dh.gov.uk/en/Publicationsandstatistics/Publications/PublicationsPolicyAndGuidance/DH_4009357.
40. Svarovsky, G. N., Shaffer, D. W. 2006. SodaConstructing an understanding of physics: Technology-based engineering activities for middle school students. *36th ASEE/IEEE Frontiers in Education Conference*, San Diego, CA. Available at http://epistemicgames.org/cv/papers/FIE_sodaconstructor_final.pdf.
41. Tlili, A., Essalmi, F., and Jemni, M. 2016. Improving learning computer architecture through an educational mobile game. *Smart Learning Environments*, 3(1), 7.
42. Vallejo, V., Mitache, A. V., Tarnanas, I., Müri, R., Mosimann, U. P., and Nef, T. 2015. Combining qualitative and quantitative methods to analyze serious games outcomes: A pilot study for a new cognitive screening tool. In *2015 37th Annual International Conference of the IEEE Engineering in Medicine and Biology Society (EMBC)*, pp. 1327–1330. IEEE., Piscataway, NJ, USA.

Index